Published by H.I. Enterprises

www.hienterprises.us

ISBN 9781734183122

Contents

Introduction

I have found that it helps, when all else fails, to holler. Hollering usually attracts attention and makes your offender want to escape. Hollering, as I have used it in the past, included complaining to organizational leaders, Inspector Generals, and Congresspersons. I have also hollered via formal complaints to the Merit Systems Protection Board (MSPB) and the Equal Employment Opportunity Commission (EEOC). The impetus for my decision to use the hollering tactic stemmed from a nursery rhyme I learned in my childhood. The rhyme went like this; Eeny meeny miny moe; Catch a nigger by the toe; If he hollers let it go; eeny meeny miny moe.

I use the metaphor "Cats Don't Bark" to remind me to accept people and situations as they are. In the metaphor, cats symbolize unethical, and sometimes immoral, people while dogs symbolize people of ethical and moral character. The "Cats Don't Bark" metaphor reminds me that immoral people should not be expected to behave morally. We all know cats do not bark—that's impossible. The "Cats Don't Bark" metaphor communicates a message that seems obvious but is not always understood.

During my 29-year federal government career, I lost numerous battles where I attempted to force cats to bark. I had to learn that I could not force people to behave the way I felt they should. Even if my way was the best/legal way, I had to accept that other paths were available for other people. My stress level increased dramatically when someone on my team knowingly took actions that were illegal. It was only when I recognized and accepted the Cats metaphor that I began to mentally and spiritually separate myself from the infectious actions of my teammates. I began to heal myself (i.e., my soul), lower my stress levels, and realize the enjoyments of life that had always stood before me. I believe if more people adopt the Cats Don't Bark metaphor and make it part of their everyday thinking, improvements in personal and professional decision-making will result and an improvement in each adopter's quality of life will be realized. I'm getting ahead of myself. Let's rewind for a moment. Let me tell you my story.

Juney

On February 18, 1966, Stevie Wonder's Uptight (Everything's Alright) was the number one Rhythm and Blues song on the chart. Life magazine published a story titled "Early to Rise, Wealthy and Wise" that described the "Today" show's Barbara Walters' daily routine. The Cincinnati Royals beat the Baltimore Bullets 118 to 114 despite Oscar Robinson scoring 44 points for the Royals. And Oh-Yeah, I was born at 11:45 am. I was two months premature and a JR baby. My father's name immediately changed from Harroll Junius Ingram to Harroll Junius Ingram, SR. I was affectionately called Juney—short for Junius.

I grew up like many other black boys in a single-parent household. My family was poor. We lived in a housing project called Academy Park where all of the houses were painted the same color—black and white. Every resident of Academy Park was poor so making fun of each other did not lead to serious arguments or fights. We all had plastic bags taped around the inside of the windows in the house to block the cold air from seeping in through the cheaply installed windows. Some families taped newspapers to the glass of each window to keep people from seeing inside the house. Curtains were not affordable. In the winter, the curtains in my house pushed away from the window most of the day as the cold air filled the plastic.

I wasn't exactly a model child although I pretended to be—especially around my mother. I listened to my mother but often did what I felt was best and earned bumps and bruises because of it. I often ditched school and yearned to reach the age of 16 so I could drop out. College was not in my future plans. Of course, mom had other plans. Mom used threats to guide me through high school. When I told her I planned to become an Auto Mechanic after graduation, she craftily explained to me that a Mechanical Engineer was just as good. They sounded similar to me. I guess mom knew what she had on her hands and was ready to provide lots of guidance for the next few years. Although I maintained my no-college mindset, mom enrolled me into college (Norfolk State University—NSU) and pushed me out the door with my sister who was starting her sophomore year at NSU.

College

I started college wondering when it was going to end. It felt like I was a high school freshman again. I continued my pattern of striving to make a "C" since that was what mom required when I was in high school in order to avoid punishment. Surprisingly, I grew to respect and actually enjoy college most of the time.

I met numerous interesting people at NSU while traversing through the electronics engineering curriculum. One gentleman who really stood out was a member of the Nation of Islam (NOI). I often listened to my NOI classmate speak about eating the proper foods and behaving respectfully. Some of his statements matched the words mom spoke for years back at home. The reinforcements led me to become a vegetarian to maintain my health as well as show respect for God's other creatures. I began to listen to the speeches of Malcolm X and gained a healthy level of determination. I also fell in love with bean pies.

I decided I would work hard to complete my studies and earn a bachelor's of science in electronics engineering. Talk of a starting pay of $30,000/year was music to my ears. To strengthen my resume, I worked part-time as an Electrician's Helper. The Master Electrician I worked for (Mr. Harris) warned me against ever working for the federal government. Mr. Harris had several contracts with the federal government and must have had some understanding of how things were done on the inside. He told me, "Don't ever work for the federal government. If you do, no one else will ever touch you." I completed my studies, received my degree, and prepared to make the money my mother and college Professors assured me I could command.

The Hunt Begins

I eventually graduated from NSU with an engineering degree and expected to have my choice of which company would get to use my talents. I was ready to be challenged and believed, as we used to say, my shit didn't stink (i.e., I was so smart that I would be successful in anything I attempted). I mailed over 200 resumes to companies of interest and only received nice not-interested responses. I eventually applied for a federal engineering position at the naval shipyard in Portsmouth Virginia and was asked to show for an interview. During the interview, the hiring official told me he did not recognize engineering degrees from Norfolk State University. Although he did not elaborate, I believed his hiring reluctance stemmed from the fact that NSU was a predominantly black university. I wondered why I was called for an interview—surely, he saw NSU in my resume. The gentleman suggested I enroll in Old Dominion University, a predominantly white university located a few miles from NSU, and complete their engineering program. His request, unbeknownst by me at the time, facilitated the aims of an old circular the Board of Navy Commissioners issued on March 17, 1817 banning employment of all enslaved and free blacks. The circular advised:

Abuses having existed in some of the Navy yards by the introduction of improper characters for improper purposes, the Board of Navy Commissioners have deemed it necessary to direct that no slaves or negroes, except under extraordinary circumstances, shall be employed in any Navy yard in the United States, and in no case without the authority from the Board of Navy Commissioners.

I angrily left the gentleman's trailer after the interview and returned to NSU to inform my former Professor of what happened. My Professor responded by telling me NSU's Engineering Department knew about the issues related to students getting hired locally. He suggested I seek positions outside of Virginia. Since I did not want to leave home, I continued my search for a local position.

I eventually received a positive response from a company called Computer Sciences Corporation in Fairfax Virginia and was asked to come in for an interview. I was not aware until I arrived that a written math test would be given as part of the interview. They demanded an ink pen be used so they could see any mistakes made. After completing the test and entering the interview room to see my results, the two gentlemen asked me to complete a math problem on the board in front of them. Although I wondered why they required me to complete the problem on the board, I chose to believe they were shocked that I was able to correctly complete the problems on paper and needed to see me repeat the feat in person. As had become normal, I did not get the job. I realized, as I had been told, I would have to consider employment outside of Virginia.

You're Hired

I applied for an Engineer position at the Washington DC Transit Authority (Subway System), received a favorable response, and traveled there for an interview. Since my sister (Freda) lived about 30 miles outside of DC, I used the time to also visit her. The night before the interview, Freda and I went to the late showing of a movie called Flatliners at a nearby theater. Flatliners was a science fiction thriller starring Julia Roberts and Kiefer Sutherland wherein medical students stopped their hearts to experience death and journey to the other side before being resuscitated. I loved the movie but hated what followed--an early wake-up time to get to DC. Ultimately, I did not get the Transit Authority job. However, since I took the time to call and say hello to an NSU classmate who lived 60 miles away from my sister, she set me up with an interview with her employer—A Navy Test and Evaluation organization in southern Maryland. Guess what? They hired me.

The Navy Test and Evaluation organization is located in a small rural town. The majority of the townspeople relied on the naval base for their livelihood. I remember stopping at a traffic light one day at the intersection of the two main roads (Great Mills Road and State Road 235) and watching a parade pass in front of me. The parade was so short there was no need for the police to manually stop traffic. Although the town was small, I enjoyed the simple living. On the weekends, I either drove a few miles to the local flea market, passing the Amish on their horse and buggy carriages, or drove to downtown DC to purchase cassette tapes, books, and bean pies from the Pyramid Book Store on Georgia Avenue. I suppose my interest in that particular book store stemmed from the information I gleaned from my NOI former NSU classmate.

My new employment position was on the E2C Hawkeye team. The E2C was a Navy aircraft used to monitor the skies for enemy craft that might harm our nation or its military. The E2C tracked enemy aircraft, as well as ground vehicles, and advised friendly aircraft of enemy positions. Although I expected a starting salary of $30K, I was only offered $17K. Although I knew the normal starting pay for a federal government Engineer was $20K, I humbly accepted the $17K offered. On my first day, I was escorted to my desk. I scanned the small secured office for another black person. Other than the one defense contractor gentleman, I saw one black female teammate who appeared to have the mindset of a different culture. After working a few weeks in that office as a Radar Tracking System Engineer, I was told by another brotha' that I was hired to fill a quota. He told me my Supervisors asked him if he knew of any black person seeking an engineering position. Luckily, I was hired before he gave my Supervisor a name. Although the revelation left me feeling demeaned, I was still happy to finally have a "real" job.

I enjoyed learning about radar tracking systems and how tweaking the aircraft's tracking software effected the craft's ability to track moving targets. Two other guys around my age were hired around the same time I was hired. We were the next generation on our team and worked well together. Some of our older coworkers seemed to be of a different work culture. One older coworker named Dan was not hospitable toward me. Dan was a higher-ranking Engineer who had a lot of E2C experience. On my first day, he referred to me as the "FNG" and laughed with his cube-mate. Of course, I wondered what FNG meant and had a thought that the "N" stood for Nigger. Dan stared at me for a few moments, appearing to know what I thought the moniker meant, before telling me it stood for Fucking New Guy. Although Dan stood-out because he took several negative actions against me, I was able to somewhat integrate with my other teammates.

The E2C team held after-work events to build camaraderie and foster teamwork. One event was held at Jack's (one of the new employees hired shortly after me) house. Jack lived on a river. The event entailed creating teams to compete in a beer-bong contest. A beer bong is essentially a keg of beer. Everyone socialized and swam in the river while the teams drank from their beer bong throughout the evening. The team that finished their keg first was deemed the winner. Although my team did not win, we had fun. I got super-tipsy and left some of my bong juice in Jack's yard. The next morning, photos circulated of me doubled-over spilling used beer. That was not the last time I saw a photo of me secretly taken by one of my coworkers.

I fell asleep at my desk one day. When I awakened, I lifted my head off of my desk, wiped my mouth, and looked around to see if anyone saw me sleeping. Although no one was around, I saw a photo of me sleeping sitting on the corner of my desk. A coworker, instead of tapping me to wake me, took the photo and left it for me to see. I guess he wanted me to know I was being watched. The photos reminded me of the words passed on to me by my elders—be careful. Knowing I was being watched was actually helpful. It caused me to watch for those who were watching me.

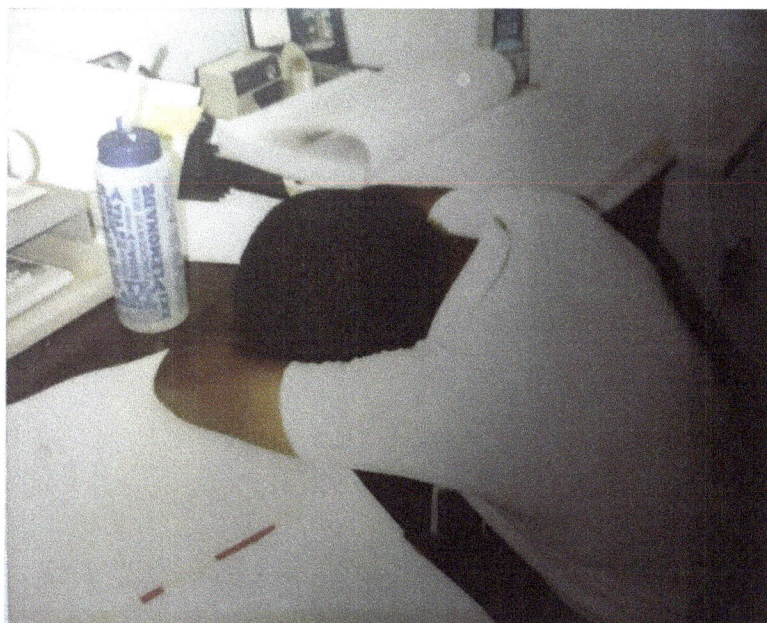

Dan had already shown interest in me. I, therefore, knew I needed to keep my eye on him if he was close in proximity. In my cube, a large glass-framed picture hung in front of me. I could see reflections behind me through the picture. Dan did not know I could see him every time he looked over the wall of my cube behind me as I typed on my computer. I knew Dan could not be trusted.

Dan and I traveled with a few other teammates for a week-long trip to New York to meet with a defense contractor. After work one evening, we all stopped by a bar for a drink and I left my drink on the bar while I went to the restroom. When I returned, Jack motioned to me indicating my drink had been touched. He then told me Dan did something to my drink while I was gone. In response, as crazy as it sounds, I stared at Dan and completed my drink in one swallow while he watched. My point was he could not harm me.

I was scheduled to travel with Dan, on another occasion, to a team meeting in New York. Since he chose to drive from Maryland to New York, I had to park my car at his house the morning we were scheduled to depart. I arrived and was invited inside by Dan's parents who were visiting from another city. As I stood in the living room being cordial with his parents, Dan entered with his two full-grown german shepherd dogs. I was not aware he had pets. Since I love dogs, I was happy to see them. The dogs were wagging their tails and sniffing me, as nice dogs do, until Dan gave the command. Dan yelled "KILL" and the dogs began to bark fiercely at me—one at each side. I froze with fear while Dan laughed. He called the dogs off and we headed out the door for our scheduled journey. On our way to New York, Dan told me his six-year-old child asked him about the differences between black kids and white kids at his school. After sharing the conversation he had with his son, Dan asked, "You do know there are differences don't you?" I told him the differences stemmed from what the children were taught by their parents. After a short discussion, we agreed to disagree and continued to our work location in New York.

I enjoyed traveling for E2C work. I felt special and empowered while traveling to different locations since the information I carried was sometimes classified. The information was locked in a double-lock container and I only had one of the two combinations—my travel-mate held the second combination. We also traveled with a letter informing anyone, including Police Officers, we could not be separated from the container. I quietly hoped a Police Officer would pull us over so we could use the letter to minimize any power he thought he had.

The classified information we traveled with had to be retrieved from a secret vault that was in another building. The keeper of that vault had a great responsibility and had to be vetted by many before given authority over the vault and its contents. It was clear he was particular about who he allowed to enter his vault. The keeper showed me he did not trust me and did not feel I had a right to enter his vault despite the fact that I had been properly cleared and had served on the team for some time—he knew me. A group of us went to the vault to get information for a flight test planned to take place in California. We trickled into the vault and I was last in line. The keeper let everyone in the vault without words until I entered. Once I entered, he yelled, "Wait! There are too many people in this vault! Everybody out!" Everyone exited the vault and re-entered. Since it all seemed strange to me, I purposely stayed out. The keeper said nothing. Obviously, there was room for everyone except me. I wonder if the vault keeper's formal security training warned against allowing people who looked like me to enter secured spaces. I seriously doubt it. I actually think his home training took precedence.

Once we reached California and arrived at the E2C pre-test meeting location, I encountered another episode similar to what I experienced at the vault. A gentleman refused to let me in the meeting with my team. He said there was not enough room for me. I explained my job as the E2C Test Engineer was to note the pre-test conditions, listen to the test plan, wait until the test event was complete, and collect the test data for evaluation. I expected the information I shared, coupled with my badge, clearance information, and his obvious knowledge that I was with the team from Maryland to cause him to step aside. I was wrong. The meeting started and he still refused to let me in. Strangely, my teammates did not lend assistance. I was forced to call my leaders back in Maryland. They contacted the gentleman's leaders in California and he was forced to let me in the meeting.

Despite the early race-based episodes at my new job, I tried hard to engage with my coworkers on a personal level and accept that I had to create a new life for myself. I mimicked what my teammates did and accepted questionable behavior towards me as how people normally behaved in the real workforce. On the inside, however, I was not satisfied. I knew I was still being treated as an outsider. I became more comfortable being by myself or with people who did not denigrate my actions simply because my actions were not what they were used to seeing. I began to decline invitations to after-work team events and started eating lunch alone or with the culturally-aligned building custodians who were also denigrated by men in suits who would leave buggers on the bathroom wall for them to remove. I began to realize the importance of accepting who I was and being true to myself.

True To Me

I declined several invitations to join my coworkers after work which was the opposite of what I had been doing. One woman eventually told me they were going to stop inviting me if I kept saying "no thank you". The invitations stopped soon after. It wasn't that I did not want to spend time with my teammates; I just wasn't really enjoying what we were doing. I knew they would not have wanted to do what interested me such as driving to the Pyramid book store on Georgia Avenue in Washington DC. I knew I had to heed the words of my mother, when she relayed the words of her I. C. Norcom High School Teacher (Ms. Wimbush), "To thine own self be true."

The Navy Test and Evaluation organization in southern Maryland was only 3.5 hours from my home in Portsmouth Virginia and I enjoyed going home to see mom every other weekend. However, I continued to have that longing for the state of Florida. I used to watch the weather report on television as a child and say one day I will live in Florida. The weather seemed to be better in Florida. The temperature seemed to always be 20 degrees warmer than Virginia. I shared my longing to live in Florida with a couple of Navy Pilots and they advised against moving to Florida due to the heat. One of the Pilots told me I would have to shower three times a day as a result of sweat. I was not deterred. After finishing my Navy-sponsored studies and earning a Master's degree in Management from Florida Tech's Maryland campus, I found an engineering job in Orlando, Florida at the Navy's Training and Simulation organization.

Living The Dream

When I arrived in Florida, I bought my first house and felt everything was headed in the right direction. I was introduced to my new teammates during my first week on the job and was happy to see several Engineers who looked like me; although they were roughly eight years older than I was. One of the older Engineers often referred to me as young-blood. He told me the group was glad to see a black Engineer had been hired after numerous years of watching new Engineers of other ethnicities hired to fill positions.

I met a brother-man 20 years my senior at a training course held at my new organization. Reganol was the Facilitator for the class and appeared to have his finger on the pulse of what was going on in the organization. After class, I sent Reganol an email message explaining that I enjoyed his class. I ended the message by asking if he knew of any black book stores in the area. My longing for more information, similar to what I received from reading books purchased at the Pyramid Book Store, led me to include the question in my email message. Reganol stopped by my desk and personally gave me the name of a book store. That action started a long friendship and mentor relationship between me and Reganol.

Although I was comfortable in my new surroundings and started to join my new coworkers in after-work activities, I soon realized I had not escaped those teammates who refused to control those urges that marginalize certain team members. My Supervisor, Wally, accused me of fraudulently making long distance calls at work. I made five calls, using my desk phone, to my house and was not aware Sanford Florida, located 20 miles away, was long distance. The organization's phone system did not warn the user when a call was considered long distance. When I explained I had to call home to make sure my girlfriend (Tracy) and her kids (Michael and Monte') were not having any issues related to the recent relocation, Wally showed no empathy and ordered me to send a $5 check to the U.S. Treasury. Soon after I sent the check, an order came down from Navy leadership informing that no one is allowed to send a check for long-distance calls to the U.S. Treasury unless the total was over $20. The cost to the taxpayer for processing checks was too high to justify processing checks below $20.

The Navy's Training and Simulation organization had an established union and I was asked to join. After realizing I might need professional protection in my new environment, I joined the union and soon after became a Steward for one of the employee divisions. As a Steward, I was tasked with researching and sharing information related to force reductions. The Navy organization was contemplating firing several of the employees in the division where I was the union Steward. I worked hard to help my people save their jobs. I eventually stepped down after realizing someone in the group was giving the organization's leaders all of the information I shared with the division employees. I was hurt that someone whose job I was trying to save was working behind my back to curry favor with management. I also noticed my leaders were beginning to treat me differently. As luck would have it, none of the employees lost his/her job.

Jackie, my new Supervisor, was not as hands-on as Wally. However, he operated secretly and with more deadly force (i.e., his actions could quickly kill a career). Jackie issued me a written reprimand that accused me of being absent without leave and time-card fraud. Jackie was fully aware I was working out of town on a mission set by the Navy employees' union with approval from the union President who informed him of the trip. To be safe, I also sent Jackie a reminder email message before I left. I was able to get the reprimand rescinded by Jackie's Supervisor after explaining the situation.

A person can spread a negative rumor about you that can result in the placement of numerous successive hurdles in your path of progression. Regardless of whether or not the rumor was true, you will be encountered by those hurdles. One brother (Jodie) told our coworkers of another culture I did not believe in God. Although he was wrong, it did not matter. That rumor spread and I believe created issues relative to my professional progression within my organization. When I questioned Jodie about what he had done, he replied that he did it because I always questioned the bible. To him, that meant I did not believe in God. I pulled my bible out and showed him the numerous highlighted sections. I told him his assessment was wrong. Another brother (Edwin), who was a party to the conversation, told Jodie he owed me an apology. Jodie responded by saying he did not owe me anything. What Jodie didn't know was I guessed he had told our coworkers I did not believe in God. I did not know he had really done that. An elder from a local bar, after I told him the story, advised me not to ever speak to Jodie again. Instead of heeding the advice, I continued to interface with Jodie. Actually, Jodie and I became good friends. Jodie was one of the most knowledgeable Engineers I met in my training systems acquisition career. Jodie obtained numerous accolades for his work and was definitely prepared for an engineering promotion. However, for unclear reasons, promotions escaped him.

Jodie and I hit many of the after-work night spots for dancing and drinks and did the things guys under age 40 do in the presence of young women. Our friendship led to Jodie requesting that I fill a short-term void he had on one of his teams at work. My Supervisor allowed me to fill-in as one of Jodie's Engineers.

I was temporarily added to Jodie's team and got my first opportunity to travel on a week-long duty assignment with another black guy. That trip turned out to be one of the best duty assignments of my career. Although the place of duty was Salt Lake City Utah, Jodie and I had a lot of fun. Don't sleep on Mormon country; surprises are there.

I later joined an interesting team that was collecting information to assist in the development of a Virginia Class submarine training simulator. We traveled to Groton Connecticut often and walked through the secured areas where the actual submarines were being built. I felt important because we had the privilege of wearing safety hats and were allowed to pass through areas where tarps covered portions of the submarines to make sure we did not see certain technologies. I knew I was pretty close to the cutting-edge of technology. I noticed I was always the only black person in the Groton team meetings. However, that did not bother me—I was used to it. I mostly sat quietly, looked, and listened. On one occasion, I heard my new teammates speak negatively about my friend, and Salt Lake City travel buddy, Jodie. Although Jodie was not a member of our team, my teammates spoke of how he often fell asleep during meetings. I suppose they were not aware of all of the medications Jodie had to take for his illnesses. Maybe they knew but did not care. Their conversations put me in an uncomfortable position. Do I tell Jodie or not? After much thought and concern for Jodie thinking those people were his friends, I mention what I heard to him. I told Jodie I wanted him to know in case there was something he could do to correct his situation. When Jodie demanded I tell him who made the comments, I refused. I knew he was going straight to my teammates and tell them what I told him. Jodie stayed angry with me for a while but everything cleared once we walked across the bridge that took us and hundreds of other partiers to Pleasure Island for dancing and drinks. Pleasure Island was a Disney property in Orlando that was filled with dance clubs and restaurants and Jodie and I were regulars. We traveled with our Pleasure Island cups that guaranteed discount refills.

My Virginia Class submarine simulation team was scheduled to fly to Groton one Monday morning and my travel arrangements were all set. The Tuesday before we were scheduled to fly out, planes flew into the twin towers in New York. The news media covered the terrorist acts all week and instilled fear in numerous people—including me. On Friday, I revealed to my team leader I was afraid of flying on Monday. She was scheduled to fly also and shared that she was afraid as well. She told me not to worry because she was going to speak with our Program Manager about delaying the trip due to the terrorist acts and would call me back. She never called back. On Monday, I realized the team made the flight without me—including my team leader. When they returned, I was told I was being replaced on the team since I could no longer fly. Although I corrected my Supervisor by telling him I could fly and that I was only afraid of taking that one trip, he did not reverse his (actually my team's) position. I was moved to another team.

On my new team, I served as Engineer under another black Lead Engineer (Mikel) who was part of our small group of disadvantaged black Engineers. I thought Mikel would lead as Jodie had and we would have a great working relationship. Although Mikel and I attended several after-work events with our culturally-aligned coworkers and had a cordial personal relationship, Mikel's professional treatment of me was very questionable. Mikel sent me to a naval base in California to install a networked classroom simulator. Mikel told me, "this is your last chance." When I asked Mikel what he meant by the statement, he replied, "if you mess this up, you will be removed from the program." I told Mikel I had never been unsuccessful in past installations.

Although I was still confused, I did not continue my query. Mikel called me to his office to make sure I understood what I needed to do in California and to make sure I had everything I needed to affect a successful installation. During our meeting, Mikel showed me two very different network cables and explained that only one was the correct cable I needed for the California installation. After explaining the differences between the cables and clearly deciding which one I needed, Mikel asked me to build the necessary cable from scratch. He added that he had all of the materials. I told Mikel I did not know how to make the cable and would like to simply take the store-bought cable he had in his hand. After going back and forth, Mikel gave me the commercially-packaged cable in his hand—the wrong cable.

One of the Sailors in California pointed out I had the wrong network cable while helping me troubleshoot why the installed simulator network was not working as planned. He told me where I could purchase the correct cable so we could successfully complete the installation. I phoned Mikel and told him of the cable issue and that I planned to purchase the correct cable and seek reimbursement after I returned to Orlando. Mikel refused to give me permission to purchase the cable. I explained to Mikel I was scheduled to leave the next morning and needed to complete the system installation as well as test the system before leaving. Mikel still refused and told me he would call me back after speaking with his Supervisor. Mikel never called me back. I contacted Mikel's Supervisor and was given permission to purchase the cable, completed the installation, and returned home for the reimbursement. I returned home expecting to be praised for the successful installation. However, Mikel was not in a praising mood. He berated me for going over his head to get permission to purchase the cable.

The confusing (i.e., negative) episodes I experienced with Jodie and Mikel taught me that the lack of an ability to reason or feel compassion for others while making important decisions was not subject to any ethnic boundaries. I never figured out why Jodie and Mikel took fratricidal actions against me that could have been detrimental to my professional progression. I had become accustomed to the behavior by the time I met Jodie and Mikel and simply added them to my list of people around whom to be cautious. Others in our small group seemed to be more supportive. For example, Reganol, the elder of our group, constantly attempted to share historical ethnic knowledge with the group in an effort to help us take stances that demanded respect. He told us we might be professionally harmed by standing for what was right but our stances would help the blacks that were hired after us—just as we were helped by the blacks before us.

I enrolled in a local college (Seminole Community College) to renew and update my computer networking skills. I needed to be able to talk-the-talk in groups of Simulation Engineers if I had any hopes of being promoted. I eventually earned an Associate's Degree in Computer Networking Technology. However, I was not able to gain a promotion while working for the Navy's Training and Simulation organization. None of the black Engineers were being promoted. Non-blacks were passing us all (i.e., being promoted to positions higher than we held) via the hiring process that existed. I applied for an Army engineering position in the building across the street and was shocked when they offered me the job with the promotion I sought. I gave my Navy Program Manager one last opportunity to keep me by asking him to match the offer. Actually, I did not want to leave. I had heard the Army followed very strict guidelines that the Navy ignored. My Program Manager responded to my final offer by saying, "Good luck in your new position." I accepted the Army's offer. I later heard Jodie, Mikel, and several other black Engineers filed a lawsuit against the Navy's Training and Simulation organization related to their discriminatory hiring practices. The lawsuit, I was told, was later withdrawn when all of the Engineers, except Jodie, were offered promotions. I suspected Jodie was the leader of the group that filed the lawsuit. Jodie was never promoted and disgracefully retired several years later.

Be All I Can Be

I was given a start date for my new Army position at the Army's Training and Simulation organization. Although my start date was a week away and I was still working for the Navy's Training and Simulation organization, one of my future Army coworkers located my Navy email address and sent me Army work to complete. I was tasked to review and edit a system specification document. I realized the earlier warning was true concerning how strict the Army was towards its employees. Of course, I researched the Army system and edited the document the Lead Engineer (Byron) sent to me. Viewing the glass as half full—versus half empty, I was able to get an early start on coming up to speed on my new project. That enabled me to hit the ground running, so to speak.

My first hint that the Army organization's hiring practices were discriminatorily subjective was quickly revealed. My new work location was in a cubical positioned between two other cubicles being used by two white guys. My Supervisor entered the cubical to the right of mine and explained to the Engineer that engineering employment positions were available and she was interested in any Engineers he knew who were interested in working for the Army. I expected her to give me the same speech after she left my coworker's cubical. However, she walked right past my cubical and entered the cubical to my left. I listened to her hiring-interest speech once again. She passed my cubical a second time on her way back to her office. I figured she was not interested in any Engineers I might refer to her for the open positions. I told several people about the positions and received about five resumes—only one was from a black person. I gave the resumes to my Supervisor but never received any updates on the filling of the engineering positions. My Supervisor's Secretary later told me, weeks before she was due to retire, my Supervisor used a scheme for hiring. The Secretary told me my Supervisor would place the resumes of the people she pre-selected (i.e., chose without a proper competition) at the top of the stack of resumes she sent to human resources. The Secretary was angry at my Supervisor for not honoring an agreement made concerning approving the Secretary's early retirement application that would have resulted in a $25,000 payment for retiring early. Although I sympathized with the Secretary, I could not help but feel she should have known better than to believe my racist Supervisor would help a black woman gain a slot that offered $25,000 and early retirement rather than give the slot to one of the white women who applied.

The Army civilian position I was hired to serve in was as an Engineer on the UH-60 Black Hawk Flight Simulator program. The assignment was interesting since the government-contractor team included a gentleman who was the subject of a published best-selling book titled Black Hawk Down and had actually served as a Pilot in one of the downed UH-60s in Mogadishu Somalia. I met the former Army Pilot at a project meeting. I had hoped I would read the book before meeting him. However, that did not happen. Actually, I never read the book. I did watch the movie and was left feeling he was a lucky man. I also felt the Soldiers who perished as a result of the helicopter crashes and subsequent clashes were the real heroes. I was happy to hear the former Army Pilot say as much in a later meeting.

I was placed in the Visual Systems Engineer position on the flight simulator team even though I had no visual systems experience. I worked hard and fast to learn enough about visual system technologies to be able to take part in discussions. I noticed the more I learned the more opinions I had to share with the team. I quickly realized my opinions were not well accepted by the team leaders. I was expected to sit quietly and learn. I was able to use my new-found knowledge to shine a light on issues I thought had gone unnoticed by the team. It turned out my team was usually aware of the issues and pretended to not recognize them for reasons not tied to the organization's mission. In one case, I showed the team the contractor's attempt to give the government a product that was cheaper and less reliable than the one they were contracted to deliver as part of the system. The contractor admitted the replacement would not fully meet the requirement. To me, it was a no-brainer that the government would decline the contractor's offer to use the cheaper product. My team leaders, however, felt I needed to stand down. I was told, "Don't fall on your sword over this." I noticed I began to be treated unfairly as a result of the revelations I shared.

The Black Hawk flight simulator Program Manager (Monica) called a team meeting via an email message and the subject of the message showed the meeting was for a status discussion. When I arrived and took a seat, Monica began to ask me direct questions concerning my position on the unreliable visual system product offered by our contractor. I took a beating that day. Nothing else was discussed at the meeting. When the meeting adjourned, a teammate (Ray) stopped me in the hallway to apologize and to tell me he had no idea Monica was going to use that meeting to attack me. He said he would never sit for another meeting like that again. Ray was a nice guy. I really appreciated the words he shared.

Monica complained to my Supervisor that I was not a team player and my Supervisor (Brad) put restrictions on my ability to send email messages. Brad gave no consideration to the information in the past messages I sent or whether or not the messages supported the organization's mission. After accepting the limitations Brad put on me for about two weeks, I told him I felt he was discriminating against me since none of his other Engineers was being held to the same limitation. Brad responded, "I am discriminating against you; just as I discriminate when I buy my milk." He lifted the restrictions a few days later.

Brad's Supervisor sent me, as the team's representative, to a meeting at Fort Rucker. The people at Fort Rucker informed me of several complaints concerning the systems we delivered to them and I took great notes to share with my team. After reading my notes, Brad's leaders, as well as the Fort Rucker representatives, praised me on a job well done. However, Monica and Brad felt I threw the team under the bus by sending my notes to leadership. Soon after, Brad decided to join me on my next assignment.

On my next travel assignment, since he was my Supervisor, Brad took control of our modes of travel (i.e., he chose the airline, rental car, and hotel). Brad knew I was a vegetarian and had been one for over ten years. He also knew my reason for being a vegetarian concerned animal rights. Regardless, as we left the airport headed to our hotel, Brad drove out of our way to visit a taxidermy store called Cabela's. I had never heard of Cabela's and thought it was just another sports store. I walked in and saw all of the murdered animals hanging around the walls. To make sure Brad did not take pleasure in his aim to harm me emotionally, I walked the whole store pretending to admire the murdered-and-stuffed beings. Brad even insisted we eat lunch in the store's cafeteria. Luckily, they offered a garden salad. We spent hours in that store. On the way out the door, I grabbed a catalog and pretended to read it as we completed our road trip. Shortly after we returned to Orlando, Brad told me I was being moved off of the UH60 Black Hawk and CH47 Chinook simulator programs per Monica's request. The team to which I was being moved was developing a driver simulator.

My new project was called the Common Driver Trainer (CDT) and the simulator being developed was planned to be used to teach Soldiers to drive Army vehicles. The first vehicle being simulated by my new team was the Stryker. The Stryker, named after two veterans who served in past wars, was designed to transport Soldiers to the battlefield quickly and securely. I was added to the team as the Hardware Engineer. Within a few months, the team's Lead Engineer (Mando) decided to leave the team. His Supervisor required him to find a replacement before he could leave so he offered me as his replacement. Mando described me as a qualified Lead Engineer better than I ever described myself. It worked; I became the new Lead Engineer for the CDT team. The other Engineers on the team were not ecstatic about me, the newest and youngest Engineer on the team, leapfrogging them and becoming their leader. This was my first leadership position. I tried to hide my feelings of ineptitude. Before Mando left the team, he spent a few minutes preparing me for my new position. He told me I would do fine but I needed to be more assertive to ensure the contractor did not "run over me" (i.e., did not take advantage of my shy and kind nature). Little did I know the project was in shambles and had hit a brick wall. My job was to get things moving again.

My forte of solving problems enabled me to motivate my team to put the CDT program back on track. Although the prime contractor appeared to not be happy with my by-the-book actions, their sub-contractor secretly praised me for putting the project back on track. My team closely monitored the development activities of the first CDT and documented issues related to requirements not being met. Although the contractor worked hard and long to correct all of the issues, time was money. The rework efforts to make corrections had a negative effect on the company's profits and resulted in complaints to my leaders, visits from the contractor's leaders, and contractor team leadership replacements. At one meeting, where I discussed the numerous test failures showing the CDT was not meeting the Army's requirements, our ground systems division leader (Lieutenant Colonel Skippy) asked me, "What the hell is your freaking problem?"

I responded, "What do you mean, sir?" To which, he didn't respond. During a follow-on meeting with Lieutenant Colonel (LTC) Skippy, he ordered me to help the contractor pass the government's tests. He ordered me to tell the contractor, in writing prior to conducting any test, every step we planned to take during the test event. He added that I had to reveal the test steps in the exact order we planned to follow and we could not deviate from the order given. LTC Skippy ordered me to deem the CDT ready for delivery and ready for Soldier-use if the contractor passed at least 10 of the 100 test scenarios we were scheduled to run. He said a 10% pass rate was successful. I whole-heartedly disagreed with him but I followed his orders. However, the contractor continued to fail a large majority of the tests. The success rate during the subject testing event was 1%. We were scheduled to present the test results to a large number of people, including Army Officers, via a teleconference. LTC Skippy said he would handle the presentation. Prior to the teleconference, an Education Specialist CDT team member (Ruby) who attended the test event voiced her negative opinions about the CDT test results. Ruby's words made our military team leader (Major D) angry and he began to yell at her while walking toward her in a threatening way. I jumped up from my seat and arrived at the commotion just in time to step between them. I kept my back to Major D while facing Ruby. Although I never told her, Ruby reminded me of my mother. My mother, at the time, was an Education Specialist for the Navy. I was surprised to see Ruby had no fear in her eyes or posture. Major D began to push my shoulders and back as if I was keeping him away from Ruby. Being from the hood and witnessing many bluffs, I could tell Major D did not want any parts of Ruby. Perhaps her non-fear posture stemmed from her knowledge of bluffs as well. I was eventually able to calm the situation. Within seconds, LTC Skippy, MAJ D's boss, stepped out of the CDT simulator located about 15 steps away from us. I did not know he was sitting in the simulator listening to the whole commotion. I lost all respect for him after that. I found it hard to accept that LTC Skippy was going to allow Major D to engage Ruby in an inappropriate way. The behavior of both Army Officers was clearly unbecoming.

LTC Skippy, although the CDT failed miserably, still tried to persuade a higher-ranking Army officer (a Colonel—COL) to allow our organization to deliver the CDT for Soldier training. I was elated to hear one of the Sergeants who assisted us with testing the CDT tell the Army COL, "Sir, if my Soldiers train in the CDT, I will have to re-train them once they get out." Although I was happy to hear the Sergeant (an expert Stryker driver) speak truth to power and admired his courage, I feared for his career. I felt he would more-than-likely be punished for speaking against LTC Skippy. As a result of the Soldier's honesty, the Colonel ordered us to make additional corrections to the CDT before delivering the simulator to the field. It didn't take long for the rumors of me being replaced as the CDT Lead Engineer and Test Director to spread.

I was told by an Army Major that LTC Skippy was about to remove me from the team because of the numerous system discrepancies I continued to document. The MAJ, a brutha', also told me he had to approach LTC Skippy about speaking to him in a disrespectful manner in front of other Officers. I walked right up to LTC Skippy and asked him if the rumor was true. LTC Skippy denied the rumor and said he was satisfied with my work. I continued to assist the contractor in correcting existing issues while documenting new failures. The CDT contractor continued working to actively reduce the existing discrepancies and the issue-total began to drop remarkably. However, the pace was not acceptable to my organization's leaders. As the Lead Engineer and Test Director, I controlled the closing of each discrepancy. My control was usurped by my Supervisory (Johnson) once I took a well-deserved week-long vacation. Johnson was asked to close a collection of the serious discrepancies I wrote. Johnson admitted to me, once I returned from my vacation, he did not understand the discrepancies he closed. Several team members who were aware of the issues and agreed the written discrepancies were valid asked me about the closures. One stated, "They waited for you to go on vacation to close the major discrepancies."

I knew there was more to leadership than what I had in my head. Although I noticed many of my coworkers who served as Lead Engineer or served in some other leadership role had mentors who guided them in their role, I had no one. My leaders did not offer to help me better understand how to lead a team. To gain leadership experience and knowledge, I joined an outside volunteer organization in Orlando Florida called the Sons Of American Veterans (SOA). The SOA is a subordinate organization to the overarching American Veterans (AMVETS) organization. AMVETS is a congressionally-chartered veterans service organization that works to improve the lives of over 20 million veteran families. The SOA works to assist the AMVET organization meet its mission. All of my SOA brothers were older than I was and offered many lessons that helped me better understand how to handle conflict and build relationships—very important leadership lessons. I served in many SOA leadership roles including several years as our Squadron-30 Commander. I used the lessons I learned in the SOA to help me better lead my CDT engineering team.

I got the hang of leading my engineering team. However, I knew my leaders were in need of training. They were leading improperly. In my view, they especially needed to be trained in leading culturally diverse teams. I was certain I could be a more effective leader than the ones in my organization. Therefore, I decided to enroll in a doctoral leadership program at the University of Phoenix. I felt I would gain the necessary leadership skills to help change the leadership culture and norms at my organization. During my studies, I completed all of my classes and chose a doctoral dissertation subject that was closely tied to the leadership challenges at my organization. The results of my 2009-published doctoral dissertation were unkind to my organization. The title of my dissertation was Organizational Transparency, Employee Perceptions, and Employee Morale: A Correlational Study.

My dissertation included a dedication to "my employer and the United States (U.S.) taxpayers" for assisting in financing the research study. I added that my organization "served as an impetus to the recognition of the problem areas researched in the current research study". I suggested further studies on the subjects included in my dissertation to reveal whether or not my findings were generalizable (i.e., could be applied to settings other than the one included in my study).

The results of my research were verified by the results of a later survey taken by a Major General who served as my organization's leader (i.e., the Program Executive Officer). The General's results showed problems of nepotism and other ethical issues at our organization. I accepted my research as generalizable when I read the results of an even later survey taken by Army headquarters where the results matched the ones in my study. My full dissertation can be found on amazon.com.

The CDT program was the Army Training and Simulation organization's flagship program while I served as the Lead Engineer. The CDT was briefed and displayed at numerous conferences. The device was named "Common" because the motion base that simulated the movements of the vehicle being simulated was designed to be used with other simulated cabs. Cabs are the part of the vehicle in which the Soldier/driver sits. The CDT was deemed "Common" because it could be used to simulate several Army vehicles. The Stryker was the first vehicle simulated. The Abrams, MRAP, and other vehicles were scheduled to be simulated using the same motion base. Costs savings were expected and advertised since Army units that purchased the CDT-Stryker would only need to purchase a new vehicle cab, the portion of the simulator that connects to the motion base, to train Soldiers to drive the other simulated vehicles. My team was given a list of five new vehicles to be designed for use with the CDT motion base. My engineering team of four people was about to grow in preparation for the new projects. For planning purposes, my team agreed on a strategy to begin working on the five new programs while we waited for additional engineering support. Unbeknown to me, LTC Skippy was finalizing plans to remove me from the CDT program.

I suspected my days on the CDT team were numbered when I heard LTC Skippy yell from a cubicle across the aisle from me, "He needs a fucking mentor!" I leaned around my cubical entrance and saw him speaking with Jimmy (one of the two people he reportedly stated were his options for replacing me on the CDT program). A few weeks later, I was replaced as the Lead Engineer on the program. Darvy, an older white Engineer was chosen as my replacement. I always wondered if Jimmy felt slighted by the selection. My Supervisor told me LTC Skippy told him he needed someone with more experience to lead the CDT program since additional vehicle projects were on the horizon. LTC Skippy added, according to Johnson, he did not believe I would be able to handle the additional responsibilities. Of course, although I had no control over the matter, I disagreed. My team was ready to start work on the additional vehicle simulators. Darvy was added to the CDT team as I was removed and he quickly closed 75% of the discrepancies I documented. Soon after, Darvy was promoted.

I sent an email message to my team informing that I was being replaced and would be moving to a different team. I started the message like I often did by saying, "Here comes Mr. change again…." After reading my message, two of my Engineers sent messages to our leaders requesting they reconsider their decision to replace me. One message was borderline disrespectful. I was really touched by my teammates' attempt to defend me. One of them sent me an email message apologizing for what our leaders had done. He was Asian and had not mastered the nuances of the English language. He told me he wept when they removed me from the team. I knew his choice of words had been gleaned from the bible. I wept after reading his message and almost every time I think about his words.

I was moved to a relatively new program called the Medical Simulation Training Center (MSTC) program. The MSTC program outfitted training centers used to train Army First Responders (Medics and Combat Life Savers) with training simulators (e.g., full-body mannequins, upper torso, IV arms) and medical equipment. The MSTC training prepared Soldiers to respond to battlefield injuries. I read the history of the MSTC project and realized the small 4-person team was formed quickly, roughly a year before I was added to the team, using war-funding dollars which allowed the team to justify streamlining the normal acquisition processes for standing-up and outfitting multiple medical training venues around the world. Although I was expected to help the team patch the holes in the acquisition processes used and give the appearance that the full acquisition process had been used, the team resented my suggestions.

My job entailed developing contractual documents necessary for acquiring medical training equipment, having the equipment delivered to numerous MSTC locations, testing the acquired equipment at each location, and ensuring the equipment was able to be maintained to extend the life of the products. I also attended numerous events where medical training needs were discussed and plans developed to satisfy those needs. A new MSTC military team leader (MAJ Dak) was added to the project as I was added. As new members of the team, MAJ Dak and I worked well together. Although I noticed he spoke ill of people (after they walked away) who disagreed with him, MAJ Dak had not shown ill-feelings toward me—at least, not in my presence. MAJ Dak mentioned he was added to the team as a result of negative feelings his previous leaders held about him. He told me he might not get promoted to LTC. I did not suspect anything when he added that we were in the same boat.

MAJ Dak was very ambitious when it came to what our little MSTC team could accomplish. He set very lofty goals for the team and many of my teammates secretly joked about MAJ Dak being delusional. I simply listened to his plans and did my best to carry them out. Because we were a small team and MAJ Dak felt a need to have several high-level positions on our organization chart, I was given a title that sounded prestigious. I went from being a Systems Engineer to the MSTC's Assistant Deputy Product Manager for New Developments. I represented the team at conferences and spent time in research and development meetings where decisions concerning funding new projects were made. I took pride in my new cubical name and title plate. I had no idea the accolades would only exist as long as I did exactly what MAJ Dak told me to do.

MAJ Dak's ideas sometimes involved questionable activities. I was aware that we, sometimes, walk the thin line between what is acceptable and unacceptable. However, I knew better than to cross a line from which I could not return on my own. I was aware help would sometimes hide when you needed it. Since I developed an aversion for asking for help, as a result of something from my childhood—I suppose, I seldom entered improper/dangerous situations where I might need someone's help to exit. I eventually found out that risk avoidance was not a trait MAJ Dak viewed as acceptable on his team. He often explained there was an "art" to acquisition—not just rules and regulations.

MAJ Dak tasked me to lead a team of government personnel from the Army's Research organization as well as their contractors, who were developing a new medical mannequin, on a data-gathering trip to one of our MSTC locations. Our objective was to gather information concerning how the existing MSTC medical mannequins were being used to train First Responders. The information I was ordered to facilitate our guest in obtaining included videos and photos of the existing MSTC mannequins and how they were used during training. I was also expected to facilitate communications between the contractors and the MSTC Instructors and Soldiers. I immediately felt uneasy with the order since I was aware that taking photos of Soldiers on an Army installation usually required permission from someone at that installation. I also knew the Soldier's names would need to be shielded from view if we were given permission to take photos and videos. I suspected the contractors that would be accompanying us might use the photos and videos of the existing MSTC training mannequins, another company's proprietary technology, to assist them in developing the new mannequin technology they would eventually attempt to sell to the government as a result of a future competition against the company whose technology they aimed to video with me as their escort. My tummy began to turn. Although I mentioned the restrictions concerning photos and videos at an Army installation, no one showed any concern. MAJ Dak said there was no issue. After a few days of dreading my upcoming task, I decided to seek information from one of my organization's Attorneys. Lisa responded and included MAJ Dak on her response message. She informed us that we could not continue with the planned activities at the MSTC if we were going to photo and video the Soldiers and/or the existing company's proprietary technologies. In response, MAJ Dak met with the travel team, informed us of the Attorney's advice, and asked the research team for a verbal promise not to photo and video the existing MSTC technology. The verbal promise was immediately given and MAJ Dak ordered us to continue with the mission. Because I was still uncomfortable and believed the "promise" request and delivery was tongue-in-cheek, I tossed and turned all night. I was aware of the research group's past actions and strongly suspected they would do whatever they wanted to do, with their contractors in tow, once I escorted them to the MSTC. I convinced myself that everything would be alright and that I should just follow orders. However, I received an email message from the Secretary of the Army concerning violating proprietary rights. The message was actually sent to all Army personnel. The message warned that anyone who violated those rights could be prosecuted. My heart filled with fear. Only a few days away from the travel date and I knew I could not lead the team in good conscience. I decided to tell MAJ Dak.

I told MAJ Dak, in writing, I could not lead the research team to our MSTC. I included my engineering Supervisor (Johnson) on the message. MAJ Dak was forced to postpone the trip since someone from our team needed to accompany the research group. Soon after, I began to notice the negative treatment from MAJ Dak, my MSTC teammates, and the research team leaders. I also realized my performance review scores did not represent my accomplishments. MAJ Dak and my teammates began to ignore and isolate me. Our group was told to relocate to a different floor in our building and MAJ Dak was responsible for selecting cubicle locations for his team members. All of my coworkers were given a new location. One gentleman told me in confidence that he was aware that MAJ Dak requested the movers to relocate every MSTC team member except me. The team that actually affects the moves was told to leave me at our old location. Because the gentleman showed me our new location with cubicles waiting to be formally selected, I chose a cubicle and affixed my nameplate. I also transported my own stuff, including my computer equipment, to my new location before my teammates' items were relocated. MAJ Dak never said a word to me about the relocation.

The negative treatment began to really bother me since I knew I had not done anything wrong and had probably assisted in the avoidance of a future lawsuit from our existing MSTC mannequin contractor. I grew tired of my teammates ignoring me when I said "good morning". I knew the ill-treatment, including issuing me lower-than-earned performance scores, was being done because I spoke truth to power. Eventually, MAJ Dak had Johnson remove me from his team. I was placed in a lower-ranking position on another team called the Engagement Skills Trainer 2000 (EST2000) team. I saw no other alternative than to file a formal complaint. Johnson met with me and tried to persuade me to change my mind. He told me he had also been given low performance review scores by his Supervisor. He warned me if I filed the complaint, no one would be my friend. He added, "but I will still be your friend." I thanked Johnson for the words and filed my first Merit Systems Protection Board (MSPB) complaint anyway. Little did I know, Johnson was right. Maybe he had seen the proverbial movie before.

I had no choice but to condemn my Supervisor's actions in my MSPB complaint. It was Johnson who issued me the low performance review scores and removed me from my leadership position on the MSTC program. Although he blamed his Supervisors and said he was just following orders, his name was all over the retaliatory actions taken against me. I felt guilty because it was Johnson who hired me from the Navy's Training and Simulation organization which resulted in a promotion. I fought back against my guilty feelings and proceeded through the complaint process. Johnson had given me several reasons to ignore my guilty feelings.

It was Johnson who closed the major discrepancies I wrote on the CDT team without my knowledge or consent. Johnson was also guilty of approving my removal from my leadership position on the CDT team and placed me in a lower-level position. Most importantly, I overheard Johnson discussing and improperly addressing a harassment complaint with his Supervisor. A young Asian coworker of mine asked Johnson to help her stop the sexual advances of a much older white male coworker whom she had repeatedly told to stop harassing her. Rather than handle the complaint in accordance with established procedures, I overheard Johnson reduce the sexual harassment to a communication gap. He pretended the well-spoken Asian woman had difficulty with the English language and had not clearly communicated her disinterest in our white coworker. I lost much respect for Johnson that day.

The EST2000 was an indoor multi-lane small-arms, crew-served, and individual anti-tank weapons training simulator. The EST2000 trained individual marksmanship and collective gunnery for dismounted Soldiers. Although I was in a subservient role to a Lead Engineer on my new program, I was given the lions-share of the work. The program had 23 ongoing projects and I was assigned responsibility for 22 of them. The Lead Engineer was responsible for only one project and he assigned one of the other Engineers on the team to help him. That left me with two newly hired Engineers to help me with my tasks. I asked the Product Manager to more evenly adjust the workload. After he refused to take any action, I sought assistance from my Supervisor. When Johnson failed to provide assistance, I sought help from a Supervisor of another group of Engineers—Darvy. Darvy was the gentleman who replaced me on the CDT program and was subsequently promoted to Supervisor. Although he had no authority over the EST2000 Engineers, Darvy called an EST2000 engineering meeting and divided the EST2000 workload. Surprisingly, the team adhered to his order without question. Johnson was not at the meeting and never said a word about Darvy stepping in his lane.

I worked the EST2000 program for a while as my MSPB complaint made its way through the legal administrative process. The EST2000 program eventually spun-off a Foreign Military Sales (FMS) arm and I was directed to serve as the lone FMS Engineer on that program. A new Program Director (Charley) was assigned to join me on the EST2000 FMS program. Charley and I worked well together as a two-man team. Our first trip was to Egypt and it was a successful venture that will always remain one of my key EST2000 memories. Charley's time on the program came to an abrupt end when his services were needed on the Army's Bradley vehicle simulator to fill the Program Director position that had been recently vacated. Also, the Lead Engineer on the team, Ray, had become gravely ill and was also leaving the Bradley team. Ray was the gentleman and former teammate who apologized to me for the way Monica treated me on the Black Hawk team.

Charley requested I be added to the Bradley team with him to fill Ray's Lead Engineer position. Johnson approved Charley's request and moved me from the EST program to the Bradley program. However, Johnson's Supervisor (Wendy), according to Johnson, added another Engineer (Tammy) from another program. As you can imagine, filling one Lead Engineer position with two Lead Engineers created a problem. When I questioned Charley and Johnson about whether or not I was added to the team as the Lead Engineer, Charley's response was yes but only for the existing Bradley projects. Charley said Tammy would Lead the new Bradley project. Johnson's response was yes also but he added that Tammy was the "Super Lead". Johnson knew there was no such position called "Super Lead". I realized Johnson had placed Tammy on the team as my oversight even though Tammy and I were the same rank. Tammy, who was known as a prima-donna on her old team, behaved as the Bradley Lead Engineer and expected me to behave as one of her Engineers.

I led the existing Bradley simulator upgrade efforts with a team of four of my organization's government employees, a support contractor, and two borrowed Army Officers (i.e., expert Bradley users). My job as the Lead Engineer, and soon to be Test Director, was to make sure the defense contractor completed the Bradley upgrade in accordance with the government's requirements. To test the contractor's work, I relied on the two Army experts with support from my government and support contractor teammates. My support contractor teammate (Randy) formally served in the Army as part of a Bradley crew and had experience with the Bradley vehicle weapons systems. Although his knowledge was considered old since the version of the Bradley he served on was no longer in use, Randy was a valued member of our team. The problem I continuously experienced with Randy stemmed from his inability to accept me as the leader of our team. He often responded to my requests that he take certain actions by saying, "The last I heard, Tammy was overall in charge". He would do whatever Tammy requested of him but opposed my requests at almost every turn. Randy would also go behind my back and offer alternate information to our customers after I gave them my position on issues. In one instance, I reported that our defense contractor was aware of a collection of issues we found, had made corrections, and my team planned to re-check the system to ensure the corrections were adequate. I finished my report by saying we were on schedule to meet our next milestone. Randy secretly contacted the customer by email and told them the system was still broken and we would not meet our scheduled milestone. One of the recipients of Randy's message informed me of his action. I smooth things over with the customer by reminding them I had never lied to them before and was not lying now—we were on schedule. I complained to Charley about Randy and nothing of substance was done to correct the problem. Randy would also schedule government team members for testing events. When I informed Randy that scheduling test events was my job as the Test Director, he complained to Charley and Tammy. Charley and Tammy sided with Randy. Since Randy was completing duties only a government person could legally complete, I complained to my and Charley's superiors. Charley and Randy were told that scheduling test events was the job of the Test Director—my job. Charley decided he would still allow Randy to schedule testing as long as Randy sent the schedule to me before allowing the team to see the schedule. I did not like Charley's resolution but I went along for the sake of the team. However, Charley's resolution was not good enough for Randy. Randy continued to complete schedules without my knowledge and share them with the team. When I questioned Randy about his violation of Charley's order, Randy told me to just show up at the site he scheduled me for and be ready to work. I realized no help was coming for my situation.

Randy's disrespectful actions and my leadership's responsive inaction led me to take matters into my own hands. When Randy showed his disdain for my academic achievements by telling me:

> Your titles do not impress me… You can call this an episode or you can cry to
>
> (Johnson)…I no longer care anymore. I will be in the office 13 December and you and I
>
> can discuss this face to face like men…or we can have another flippin meeting where

you can whine and complain some more to everybody about how out of control I am.

Set the date and time…my schedule is on the team calendar…I will be there.

I sent Randy's message up my chain of command and stressed Randy's invitation to meet me face-to-face "like men". However, nothing of substance was done. I was simply told by my Supervisors to meet with Randy one-on-one. I responded to Randy's email message and acknowledged that my leaders refused to assist me in stopping his disrespectful behavior towards me. I added that I would send a request directly to his company leaders if he continued the behavior that was unbecoming of a government support contractor. Randy's actions actually violated his company's contract with the government. Randy responded by telling me he would forward my message "for me" to his company's CEO. He added that nothing was going to happen to him. He then forwarded the message to his CEO. He was right; nothing happened.

Randy's email message indicated he knew who his Supervisor was. However, while under oath during his deposition testimony concerning the EEO complaint I eventually filed, he said he did not know who his Supervisor was. He testified his performance evaluations were completed by his peers who had no idea what he did for my organization. My Supervisor (Johnson) later testified he did not know who Randy's Supervisor was. There was no wonder why Randy was emboldened to harass me the way he did. There was no real oversight of his actions as an Army support contractor.

My repeated pleas for help fell on deaf ears. Despite the hostile work environment created by Randy and facilitated by my leaders, my team was very successful in upgrading and delivering the Bradley simulators. Although the results of my leadership on the Bradley team led our customer to inform us that the simulator we delivered was the best Bradley training system my organization ever delivered to them, Charley and my other leaders decided to remove me from my leadership positions on the Bradley team rather than effectively address Randy's behavior. My next stop looked very familiar as I was sent back to the EST2000 team.

My leaders' decision to let a support contractor run amok in 2009 should not have come as a surprise. My organization was well known for lacking in providing oversight of its contractors. My organization's leaders were summoned to testify before the 111th Congress (Senate Armed Services Committee) concerning a lack of oversight of their contractors working in Afghanistan. The lack of oversight resulted in serious violations of government policies and the injury and death of several people. The Army's Training and Simulation organization awarded a contract to a contracting team that included employees from a company called Blackwater. Several of those contracted employees, although prohibited from carrying weapons, took weapons from a location in Afghanistan and shot and killed multiple people. The weapons violations and civilian killings were reported to my Army Training and Simulation organization. However, my organization, according to senate testimony, failed to take any timely actions. In fact, my organization proceeded to give the contractor team a bonus. According to the Department of Justice, the shootings impacted the national security interests of the United States.

The EST2000 mission and team make-up had not changed much since I last served on the team. The same Lead Engineer (Paul) was still in his position and I was again serving as one of his Engineers. As always, I provided full support as if my organization had not mistreated me in the past. Paul tasked me to lead several efforts on his team such as ensuring the defense contractor properly developed and tested the simulated weapons before we delivered EST2000 systems to Army posts around the country. The team was still developing and delivering systems to foreign countries. In one case, I was tasked to verify the contractor packaged all of the system spare parts the government paid for before the shipment was sent to one of the foreign locations. For the verification, I used a parts list sent by the contractor that included part numbers and prices the contractor charged the government. Because I was suspicious about the identifications the contractor entered for a couple of items, I closely monitored the items as they were placed in each shipping container. One item had a description that sounded like a tube of adhesive but included a $500 price tag. I was anxious to see that item. When I laid my eyes on the item, I realized it was a small tube of Loctite that sold for $5 at Home Depot. I mentioned the discrepancy to the contractor and reported my findings to Paul, our Contracts Specialist, and other leaders at my organization. Despite my finding and repeated reminders, nothing was done about the two $5 bottles of Loctite we bought for $1000 at the taxpayers' expense.

Leadership's treatment of me was beginning to look like it was organization-wide. Regardless of which team I was moved to, I was likely going to have to deal with professional (and maybe personal) disrespect. The inaction of my leaders in response to the way Randy treated me on the Bradley team left me feeling racism might have been partially responsible. Although I was no longer on the Bradley team, I decided to file my first formal EEO Complaint.

It wasn't long after I filed my first EEO complaint that the MSPB issued a response concerning my MSTC complaint. I was relieved to find they sided with me and decided I had been unlawfully removed from my Lead Engineer position by my leaders and had been unfairly scored on my annual performance review. The MSPB ordered my organization to return me to my leadership position on the MSTC team and increase my performance review score. I was headed home—so to speak. Little did I know the proverbial welcome-mat had been flipped over.

I returned to my Lead Engineer position on the MSTC team and saw a slew of new faces. Since the team was forced to replace their current Lead Engineer with me, as the MSPB ordered, not all of the faces were smiling—especially the Lead Engineer (Sally) I replaced. I idiotically expected Sally to understand why she was being displaced and just continue to support the team. I had far more experience than she had and was certain she had seen, and likely used, my past work/documents concerning the MSTC projects. Needless to say, my expectation never came to fruition. Not only did Sally not work well with me, she also refused to sit beside me in meetings. I would come into the room and sit beside her and she would get up and move to another seat on the other side of the room. I would see her talking in small groups with our teammates and wonder if I was the topic of the discussion. Eventually, Sally was moved to a new team. I was later told she was very upset about being displaced and had cried in an attempt to be returned to my position.

Other MSTC team members began to mistreat and disrespect me also. Even the team leader (LTC Reese) who replaced MAJ Dak while I was gone joined in the "fun". Before MAJ Dak had me removed from his team, I spoke to LTC Reese at a conference. When I explained to LTC Reese that I probably would not be on the team when he arrived due to a plan to move me, he emphatically stated he wanted me to remain on the team and serve as his Lead Engineer. He was so believable that I used his name and statements to try to get Johnson and MAJ Dak to allow me to remain on the team to assist LTC Reese. Imagine how surprised I was to experience the almost unbearable level of disrespect from LTC Reese once I was returned to the team. Although his negative treatment topped all other teammates, it wasn't much higher than what I experienced from the new Project Director (Fendi) who started on the same day I was returned to the team.

LTC Reese appeared to be a very focused and determined leader and was very well-spoken. I was happy to finally join his team and was eager to get back to work serving the training needs of Army First Responders in the way LTC Reese envisioned. Based on his words to me at the conference where we initially met, I expected us to make great strides in meeting the goals of the

MSTC program. Since I had served in multiple capacities on the team under MAJ Dak, I believed I would easily come up-to-speed and gel with the existing members of LTC Reese's team. My first tasks included interfacing with team members, writing work statements and other contractual documents, and assisting the team with preparations for our annual modeling and simulation conference titled Interservice/Industry Training, Simulation and Education Conference (I/ITSEC).

I quickly recognized my teammates were not accepting me as an equal team member and definitely were not interacting with me as the team's Lead Engineer in the normal way. My strategies for completing acquisition tasks that were in my lane were second-guessed and changed without my knowledge. Being new to the current team, I felt compelled to go along. The normal ability for the Lead Engineer to directly request and receive assistance from the support contractors on the team did not exist for me. I, as a government employee, was expected to take orders from the support contractors on the team. My requests for help from LTC Reese and my engineering Supervisor (Aron) were ignored.

Aron was actually the engineering Supervisor for a different division that did not include the MSTC team. I was added to Aron's list of Engineers as one of his additional duties since the MSTC no longer had an engineering Supervisor. When I met with Aron, he explained that he was not sure why he was chosen to supervise me. He added that he knew I had been on the team for a long time and had led other teams in the past. Therefore, he said he would not really be supervising me. He said I would be supervising myself but he was available if I needed his help. I was pleased that Aron was open and honest with me and that he recognized what our other organizational leaders apparently did not—I deserved a promotion to a supervisory role.

The week-long I/ITSEC was nearing and LTC Reese planned to allow his team to reserve hotel rooms in the area of the conference. He told the team he cared about our safety and did not want us to attempt to drive home each night after working long hours at the conference. All team members were given a hotel room reservation except me. When LTC Reese's team work schedule was released, I noticed I was scheduled to work extended hours every day of the week. When LTC Reese refused to allow me to reserve a hotel room like all of my teammates and refused to make any corrections to my conference work schedule, I sought help from Aron. Aron responded that he did not understand why I was scheduled for so many hours since conference booths were not open for all of the hours I was scheduled. He told me to speak with LTC Reese again. Because LTC Reese would not change any of the plans, I decided to change my own plans. I used the bus transportation my organization provided for the conference which meant I could not stay at the conference late—as scheduled. I also did not help with the cleanup duties on the last day of the conference since that would have resulted in me missing the bus transportation back to my organization. LTC Reese admonished me for not assisting with the cleanup activities.

My teammates were clearly not viewing me as a member of their team. In some cases, they appeared to act as saboteurs. One of the support contractors who was scheduled to assist me in setting up our booth and our training aids for a medical conference at a Walt Disney property in Orlando failed to show up for the assignment. Luckily, I brought Clara, my girlfriend, to the setup event and she helped me carry the heavy manikin into the building. While setting up the booth, I asked the delivery team, support contractors who delivered our equipment for all of our conferences and who had become knowledgeable concerning how our booths and manikins were assembled and emplaced, a couple of questions about booth item placements. One of the two women, who was close friends with the support contractor gentleman who failed to show up, disrespectfully responded "you have not had that AHA moment (meaning I did not understand my job). If Clara had been in the room at the time, that contractor woman would have probably been knocked on her behind. Being mild-mannered and wearing my professional hat, I ignored her snide remark. At a different conference where I was scheduled to speak to a room full of attendees, that same woman assisted the gentleman who failed to show for the Disney conference in his attempts to distract me while I was presenting. Admittedly, I was nervous and it was probably obvious to the audience. The contractor gentleman started jumping up and down in the back of the room while the woman laughed uncontrollably. Their attempts to distract me and make me commit errors worked. I repeatedly pressed the slide changer button on my hand-held prompter by mistake. The result was the slide I was showing did not match my words. Luckily, someone in the control room continuously made corrections that put me back on the proper slide. The actions of my teammates and the team leader were fratricidal—in my opinion.

LTC Reese assigned me the duty of representing our team on a teleconference one afternoon. I called into the conference and, although I recognized LTC Reese's voice on the line, I listened to the entire conference. Once the teleconference discussions ended, the conference leader informed the attendees that the remaining time would be spent discussing the one item someone mentioned earlier that was delayed until the end since it was not relevant to the teleconference subject. Although I had used the restroom just before the two-hour teleconference started because I knew I would find it difficult to not take a restroom break before the teleconference ended, I was shaking my legs trying to successfully hear the last words of the teleconference before racing to the restroom. Nature led me to hang-up and head for the restroom. When I returned, LTC Reese was at my desk. He said, "you're back?" Realizing he felt I had not listened to the full teleconference, I told him I had listened to the teleconference but had to step away to use the restroom. I offered to send him my notes and he accepted my offer. I sent LTC Reese my notes and left the office for the day. On my way home, LTC Reese called me on my work cell phone and told me he expects me to follow his order (i.e., listen to the entire teleconference) next time. I responded by reminding LTC Reese I had listened to the full discussions but really had to use the restroom. In an effort to help him understand the urgency I felt at the time, I told him I had to urinate and defecate. That revelation led him to request my Supervisor issue me a Letter-Of-Caution for telling him I had to urinate and defecate. Aron and his Supervisor quickly complied and wrote that my words were "unprofessional, indecent, disrespectful, and discourteous" and unbecoming of a federal employee. The letter also stated my response was "unacceptable and does not represent the proper conduct of a federal employee" and could "hinder the execution of the mission of PEO STRI". I was shocked that my use of clinical terms to a medical officer was viewed as indecent. Especially since I witnessed my teammates, on numerous occasions, curse and truly use indecent language in the presence of LTC Reese and I watched him respond with laughter.

LTC Reese, on another occasion, asked me to send him an email message showing the date I rejoined the team and any days I took off since returning to the team. He told me he needed the information for engineering support considerations. Although I was baffled by his request, I complied. In my email message, I asked LTC Reese if he had any issues with my support to the team. However, LTC Reese never responded. LTC Reese also told me I needed to integrate with the team more. When I asked LTC Reese to tell me the specific things I needed to work on he said he would send me something in an email. However, he never did. Additionally, LTC Reese gave my engineering duties to our support contractors rather than allow me to execute my technical duties as I had been trained to do for the Army and taxpayers. Of course, I informed Aron and got no response.

Fendi served as the new MSTC Project Director and I, as the only MSTC Engineer, served as her team's technical representative. Although Fendi was responsible for the technical, schedule, and cost aspects of her projects, I was responsible for advising her on the technical aspects of her projects. My job was to make sure any company we entered into a contract with to provide MSTC medical simulation products was technically capable of performing as required. Since Fendi was new to the team, I did not expect her to adopt the behaviors of the seasoned people who knew of my past MSPB whistleblower and EEO complaints. My expectations were again unfounded. I soon realized Fendi had drunk the Kool-Aid—as the saying goes. She became the new face of the group taking retaliatory actions against me.

It was clear Fendi did not want to work with me as her Engineer. She often completed my engineering duties herself, with her limited technical and/or acquisition engineering knowledge, or asked another Engineer from a different team to complete my responsibilities. Fendi completed one of my statement of work documents and sent it to another agency for action. When someone from that agency asked her to explain her document, she directed them to me for an explanation. Of course, I was not able to explain Fendi's document since I did not understand it either. The other agency, without a clear document, was not able to accept our work or take action on our request.

Fendi approved engineering documents submitted by a defense contractor without allowing me to review the documents first. When I asked her to send me one of the documents for review, she replied she would send it to me the following year. She knew that would be too late for the government to request any changes to the document from the defense contractor. Fendi also sent untrained and inexperienced organizational support contractors, assigned to our team, to sites to inspect delivered items that I should have been sent to inspect. Those people accepted substandard items as part of the deliveries and the defense contractor was paid for those items.

Fendi doctored several of my engineering documents, signed her name to the resulting documents, and sent them to our contracting division for processing. Eventually, a Contract Specialist informed me of what she was doing, asked me to correct her mistakes in the documents, and showed me how to sign and lock my documents to make sure the problem ceased.

I reported to Fendi and her Supervisors that one company, who submitted a product for a contract competition and won, actually delivered a different product than what they submitted for the competition. Since I was the technical leader for the competition and evaluated all of the submitted products, I was fully aware of which product won the competition. I added that I did not believe the delivered item would have won the competition. Fendi decided against correcting the problem so the company was paid the amount the government agreed to pay for the product that was submitted for the competition even though the product delivered was of lower quality.

I served as the technical leader for another competition on Fendi's team. Our customer, according to Fendi, pre-decided which company he wanted to win the competition. Despite my repeated efforts to make sure the technical portion of the competition was fair, our customer representative, who Fendi approved to join me on the technical evaluation team with one of his Soldiers, took actions that I felt were questionable. Fendi selected a third person, a Navy Officer, to round-out my four-person technical evaluation team. The proposal evaluation activity proved to be the toughest I had ever experienced due to the evaluation team members Fendi selected. Although my organization's policy required all evaluation team members to be trained by our evaluation experts, Fendi did not require her selected personnel (i.e., my team members) to take the training. During the technical evaluation process, the three military personnel made several attempts to usurp my authority, played loud rock music, and repeatedly played a Mr. Bojangles video while commenting on the dancer's (Mr. Bill Bojangles') movements and hair. Regardless of the distractions, I met my responsibility for submitting a final technical report that was clear and fair. However, the report had to pass through Fendi before going to our contracting division. Fendi made several changes to my report that resulted in changes to the outcome of the competition. As a result, the company that placed second filed a successful protest and we were forced to repeat the competition. Not much changed the second time around. In fact, when the evaluation team met to finalize the results, one of the military Officers commented, "why don't we just change the date on the final report we submitted the first time and resubmit it?" Although we did not accept his suggestion, once we submitted our final report the same company protested again. We were again told by the Government Accountability Office to redo the competition. The third time around I was removed as the technical lead. Fendi replaced me with the Navy Officer who had served as one of my team members during the earlier competitions. The Officer did not work for my organization and was the person who repeatedly played the Mr. Bojangles video during the initial evaluation activities. Since my replacement was part of my original team, I was aware of where he stood concerning the competition. Needless to say, the third time was the charm. Our customer got the company he wanted. My gut told me Fendi blamed me for the first two unsuccessful attempts. By the way, the winning company was the same company who was scheduled to accompany me to the MSTC to photo and video the existing company's technology—the event that led to my first MSPB complaint. The replacement Engineer Fendi used for the third competition worked for the Army research group that was scheduled to accompany me to the MSTC with the company that won the competition. The company whose product was the subject of the targeted photo and video actions was a subsidiary of the company that eventually lost the competition—the company that twice successfully protested the competition.

Fendi, as time went on, increasingly circumvented my authority, began to complain about me to my Supervisor—Aron, and asked Aron to have me removed from the MSTC program. Within 24 hours of Fendi's request, Aron questioned our human resources and legal offices on how to remove me without violating the MSPB's order that returned me to the Lead Engineer position on the MSTC team. I was under attack and did not yet know it. Proof of the aforementioned Fendi-Aron attack was eventually revealed as a result of an evidentiary discovery request during one of my MSPB cases.

I was told by many people to watch my back on the MSTC team. Apparently, they knew some of my teammates and leaders were trying to harm me and remove me, again, from my leadership position. Despite the warnings, attacks, and disrespect, I chose to stay on the team because I did not want to switch to a team that supported the killing of innocent people. I knew my past contributions on training simulators like the Bradley, Stryker, and Black Hawk were being used to facilitate killing (i.e., used to train Soldiers to kill). I also knew my government had killed many innocent people, including women and children, in the past using Soldiers who trained on our simulators. Therefore, I continued to fight for my right to serve on the MSTC team. I continued to seek fair and equal treatment from my organization's leaders who were emplaced to serve the same service members and taxpayers I was hired to serve. My decision to continue fighting led me to file a second MSPB complaint. This time, against Fendi, LTC Reese, and Aron.

The unfair treatment didn't stop with my project team. My extended coworkers also took advantage of opportunities to mistreat me. It was as if they smelled blood from the initial attacks. Some of the cats served in positions far removed from the activities of my project team. My organization's Public Affairs Officer (PAO) showed she could not control her racist urges. In an effort to promote comradery, our PAO brilliantly decided to post photos (with captions) of employees, and their family members, sharing significant accomplishments. The PAO's posts showed on the computer screens of all employees for a couple of days each month.

One month our PAO asked the employees to submit photos and captions for posting consideration. After reading her repeated requests for photos of accomplishments, I responded by sending a photo of me, my sister, and my mother in our doctoral graduation regalia. My sister had recently graduated, earning a doctorate degree, and my mother and I had also earned doctorate degrees a few years prior. The three of us took a family photo dressed in our regalia at my sister's graduation. Since the PAO's requests stopped after I submitted my family photo, I waited to see my family's graduation photo on my computer screen. I never saw the photo. The PAO, apparently, decided we would not have any accomplishments for that month.

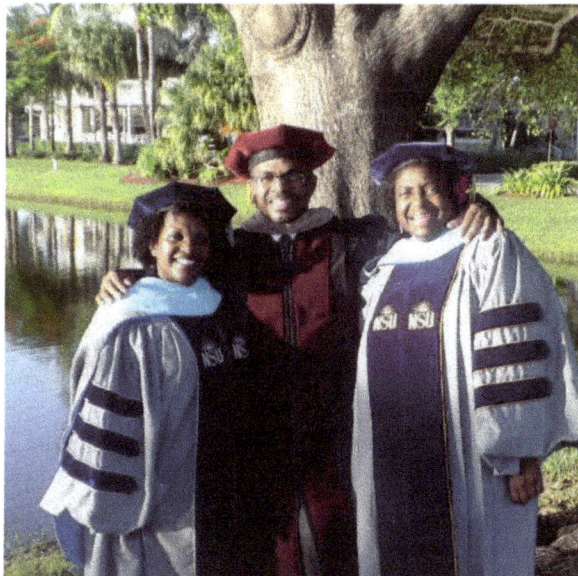

Three Ring Circus

I was raised to believe hard work would lead to rewards such as owning a house, being able to pay my bills on time, and having a good retirement that would carry me through old age. I was also taught to tell the truth and treat people like I wanted to be treated. On the professional side, I was taught to look, listen, and follow directions. I was also taught to maximize my education to increase my earnings. Therefore, I made decisions and took actions early in my career to set me up for success—I checked the right boxes. After earning my Bachelor's degree and gaining professional employment, I continued my education (i.e., I earned three additional degrees) to help me enhance my knowledge and gain employment promotions. I also earned a Lean Six Sigma certification and other certifications closely related to my employment positions. However, the federal government, at least in the places where I worked, did not recognize any of my accolades beyond my undergraduate degree. My leaders always offered advice on additional things I needed to do to earn promotions. However, after following the advice I was given, my leaders would overlook me and create additional promotion requirements. I began to feel like a dog jumping through circus hoops in hopes of getting a reward. After 29 years of work and accomplishments, I was still in a position I would have held had I stopped achieving after I earned my Bachelor's degree. I wondered if there were others suffering the same results.

I looked around to see how many blacks in my organization had been promoted to a leadership position. For over a decade we only had one black civilian who was promoted to a leadership position. In engineering, no blacks had been promoted to a leadership position. Even at the lower levels (i.e., considering the number of rank-and-file black Engineers), the number of black Engineers at my organization was less than 4% of the total number of Engineers. I was able to get the list of Engineers in the organization as part of an EEO evidentiary discovery request. It became apparent to me that my organizational leaders had no intention of promoting a black man to an engineering position where he would truly lead a multicultural team—especially if the team included white members. My organization's efforts to suppress the number of blacks in leadership positions opposed the efforts of congress to correct the underrepresentation of minorities in civilian positions within the federal government.

The federal government, the nation's largest employer, employs about three million civilians. Less than 12 % of those civilians are African American. The policy of the United States requires the federal workforce to reflect the diversity of the nation. The Civil Service Reform Act states that a properly administered federal personnel management system should provide the people of the United States with a workforce that is reflective of the Nation's diversity—federal recruitment should endeavor to achieve a workforce from all segments of society. Congress, to assist in meeting the diversity objective, placed affirmative recruitment requirements into the Reform Act. Title 5 of the US Code, section 7201, requires the Office of Personnel Management to oversee agency recruitment efforts designed to eliminate minority underrepresentation. Underrepresentation is deemed to exist, and require affirmative recruitment efforts, whenever the percentage of a minority group in any position is lower than the percentage of that minority group in the civilian labor force as a whole. Therefore, when African Americans represent 12% of the civilian workforce, having a mere 4% representation rate for African American Engineers at my organization was evidence of underrepresentation. Clearly, the EEOC should have agreed with the disparate impact claim I raised in my complaints and should have ordered my organization to take affirmative recruitment actions to correct the underrepresentation problem. Instead, the EEOC read my complaints and fell silent on the issue.

New Leadership

Almost all of the leaders I worked with since starting my professional career treated me differently, negatively speaking, than they treated my coworkers who did not look like me. It seemed I was on an unending quest for a fair and ethical leader. I admit I have had some good leaders, but not many. I wondered if my next leader would be a good one. LTC Reese's time at my organization had ended and a new leader was inbound. I conducted my usual internet search for information and photos concerning the incoming leader in hopes that I might be able to decipher his character via the search results. My gut feeling, as I evaluated my findings, was that LTC Christ would be like the past Army Officers for which I worked. He would make decisions based on how it might affect his future promotions. Was I right? Continue reading and decide for yourself.

LTC Christ started with the usual glad-to-be-a-part-of-the-team speech. He told us he had an open-door policy. His words were great. However, his eventual actions stood separately. LTC Christ brought Sally back to the MSTC team and gave her one of my engineering projects. The project was almost completed. He also sought and found another Engineer (Nettie) to add to the team as my Supervisor. Nettie had serious issues with her people-skills on the team LTC Christ rescued her from. The rumor was she was forced off of that team. LTC Christ gave Nettie her first supervisory position even though she had no supervisory training. His action upset me since I had been waiting for years to compete for that supervisory position once it was advertised. LTC Christ's gift to Nettie meant the position would not be advertised. It wasn't long before Nettie, with support from LTC Christ, began to show signs that she had accepted the baton from the past leaders who retaliated against me because of my past disclosures—she drank the Kool-Aid. My performance review scores began to drop below average with no justification (i.e., with only positive comments written in my appraisal by Nettie). The actions revealed Nettie's lack of supervisory training and experience.

LTC Christ had already shown signs of supporting his employees who refused to control their racist urges. Bertha, the team leader of the Army's research command located a block away from our location, continued her unprofessional actions against me by asking LTC Christ to remove me as the point of contact and leader of our medical simulation research projects. LTC Christ had assigned me the leadership task which required me to meet with the research command's project leaders monthly to gain updates on the three projects we paid them to complete. Bertha was the project leaders' Supervisor. Because she could not accept me as a leader, she informed LTC Christ that her people did not have time to meet with me for one hour each month to give me updates. I did not view the meeting as demanding since I only required an hour for the combined updates. I also told the project leaders they could email me their updates if they could not attend the meeting. After three months (i.e., three meetings), LTC Christ told me he no longer needed me to lead the effort. Without the monthly updates, the funds were depleted by the research command without oversight and we never received the final products we required and paid for with taxpayers' dollars.

Bertha's disrespectful actions toward me preceded LTC Christ's tenure on our team. I served as my team's representative on a medical consortium where representatives from several organizations met to discuss ideas for future medical simulation technologies. The Army's research command hosted the meetings and Bertha served as the host. In one meeting the participants, including Bertha, shared their thoughts on a particular subject. Once I began to speak, Bertha interrupted me and told the participants it was a good time to take a break. Surprisingly, although I expected someone to intervene on my behalf, the room emptied. I was the only person who remained in the room. Once everyone returned, Bertha moved on to the next subject as if I had not spoken a word. I wondered if my physical attributes played a role in the contagious decision to be disrespectful. No one in the room looked like me.

Nettie was a good friend of Fendi and Fendi was a good friend of our past leader, LTC Reese. They were all from the same commonwealth—Puerto Rico. We also had a Puerto Rican Contracting Officer and Logistician on the team. So, I was not totally surprised by Nettie's behavior toward me. Fendi wanted me off the team and I suspected she would work though Nettie, as she attempted to do with my past Supervisor—Aron, to get what she wanted. Although I had no intention of being forced off the team easily, I knew I had no one to turn to once the small group, secretly known by several people as the Puerto Rican Mafia, started attacking me. The attacks came in waves. I realized I developed a mistrust of Puerto Ricans. Once someone told me they were from Puerto Rico, my introverted personality would take over as a mechanism of protection. After much thought, I realized my reaction was no better than the woman who clutches her purse as I walk on an elevator. The elevator woman and I judged the masses by the actions of a few.

I served as the Engineer on the MSTC's Medical Training Evaluation System (MTES) project team. The team's mission was to acquire, develop, and deliver a system that could be used to assess student training as well as assess the quality of the Instructor-led training at the 21 MSTCs around the world. Fendi served as the project team's leader. We completed the normal selection process to choose a small business to develop and install the MTES devices. The winning small business contractor (i.e., the prime contractor) selected two sub-contractors to assistance them. One of the sub-contractors had a known history of developing the type of software necessary for the MTES. I had worked with them in the past on an MTES-like development effort. The other sub-contractor was not very familiar to me.

I had met the owner of the second sub-contractor at a conference a few months prior to the MTES contract award and he told me he was very good friends with LTC Reese. Although they were both Puerto Ricans, I did not assume the owner of the company was cut from the same cloth as LTC Reese. He turned out to be a nice guy. When I asked LTC Reese if they were good friends, he quickly responded that the friendship statement was not true. I remember feeling LTC Reese was lying again—something he often did.

Once the MTES work started, I was surprised to hear the Puerto Rican sub-contractor speak negatively about the prime MTES contractor. Especially since the prime contractor was his boss and could remove his company from the team at any time. The sub-contractor told us his boss was overcharging us for the MTES work. I questioned myself concerning my feelings about the information shared. The information was good for the government to know. So why was I bothered by the sharing of the information? Was it because the prime contractor was a black-owned company?

We made the MTES contract a multi-year contract because we knew the requirements would change periodically and we needed flexibility so we could require the contractor to update the systems annually. The first year went well and we were making plans to start the second year. We held the initial second year meeting to make sure the contractor was aware of the required changes to the systems and to make sure everyone was in agreement with the costs and schedule. During the meeting, Fendi asked the prime contractor if he intended to keep the Puerto Rican company as a sub-contractor. The prime contractor informed that they no longer needed that sub-contractor for the second year. I was puzzled by Fendi's question and wondered why she asked it. The government usually interfaces with the prime contractor on efforts. We usually do not question the prime about his sub-contractor decisions after the contract has been awarded. Soon after we were informed the sub-contractor was no longer part of the MTES team, things really changed concerning the MTES project. For starters, I was given the helm. I was told Fendi would be available if I needed any program management assistance. Although up for the challenge, I was an Engineer—not a Program Manager. However, I was happily shocked.

I had already obtained a certification in program management from the DODs Acquisition University. I had also completed program management tasks in the past and had actually handled most of the MTES actions during the first year. I was happy to be given the challenge of serving as the Engineer and Program Manager for the MTES project. But I was not aware that I would lose all of my team support and would have to virtually play every position on the government's MTES team. I sat as the only person from the government in 10 of the 12 monthly meetings during the second year. During the first year, we had no less than five or six people at the table for each meeting. The MTES prime contractor eventually started questioning me on why no one else attended the meetings. They asked if they had done anything wrong. The running question was, "how come they don't love us anymore?" They also told me they, after requesting numerous times, had not been granted a meeting with our new leader LTC Christ. I was embarrassed because I knew LTC Christ normally made it a point to meet with all of our contracted companies. I had no explanations for my team's actions—inactions.

I and the MTES prime contractor soon realized we needed to rely on each other for mutual success. We recognized we were a group of blacks, and three honorary blacks who worked for the contractor, meeting monthly trying to make sure we were successful so my government team would continue the contract for the third year. I needed to be successful for additional reasons.

Someone from the prime contractor's team shared with me that his company was forced to hire the Puerto Rican sub-contractor to win the initial contract. The government's original plan, I was later told by a government coworker, was to award the MTES contract to the Puerto Rican-owned company rather than the black-owned company. However, the Puerto Rican-owned company was not yet qualified to serve as the prime contractor. Their "paperwork" was not in order, I was told. The information served as missing puzzle pieces for me. The information helped me understand why my teammates virtually stopped working the MTES effort after it was clear the Puerto Rican-owned company would be fired from the project.

The MTES project was progressing almost perfectly and I had not asked for assistance from Fendi or any other leader on my government team. I briefed my government team, weekly, on the progress of the MTES project and took notice of the scrutinizing questions I had to answer. I realized my teammates took a special interest whenever I mentioned a possible issue on the project. The two monthly MTES meetings they chose to attend during the second year were meetings that occurred after I informed my government teammates of a possible issue. I made sure none of the possible issues came to fruition and enjoyed the looks of disappointment from certain government teammates. The MTES team donned happy-faces due to our long-running success and the fact that we were weeks away from starting the third year.

Year three was set to be a remarkable year for MTES. Our MTES was about to be illuminated for the entire Army medical community to see and access remotely. I was certain the MTES contractor team would earn accolades and additional contracts as a result of our work. My only issue was gaining the necessary approvals from my team leaders to start the third year. Of course, my leaders delayed approval without giving a reason. Once the approval date came and went, I asked our Contracting Officer, a believed member of that proverbial Puerto Rican mafia, when he planned to sign the contract for year three. He told me to wait two more weeks—again no reason was given. I reported his answer to the MTES prime contractor who needed assurance of the third-year contract for financial purposes since they were a small business. Surprisingly, the owner of the company started becoming difficult and disagreeable. At one point, when I responded to his threat of not completing a requirement by informing him that the action would end the contract and there would be no year three, he replied by saying maybe his company did not need a third year. I was baffled by his response and due to leave for the day in 30 minutes. I told him I would have to report his position to my team once I left the meeting and returned to my office. I reminded him I worked for the government. I was happy he relented just before I left. I noticed a difference in how the owner interfaced with me as the government's representative during year two of the contract compared to how he interfaced with the non-black representatives during year one. I overlooked my feelings since I did not have to work directly with him. I had no issues with his deputies and his on-the-ground workers. I suppose he felt slighted by my team leaders when they sent lowly me as the government's MTES leader. On the bright side, we were down to one week before signature-time to start the third year.

It was a Monday morning when I got the bad news. This was supposed to be the week we obtained the Contracting Officer's signature to start our planned MTES work for year-three. Our first travel location was set and financing was in place. After I briefed the government team on the status of MTES, the Contracting Officer told the team his office received a call from the MTES prime contractor's owner over the weekend saying they needed more funding to complete the year-two work. I was confused by the revelation since we were down to minimal year-two closeout activities that were covered by the remaining funds on the contract. Although I informed the team there must be a mistake and that I would resolve it, the Contracting Officer said he wanted a full audit. He also decided to further delay signing the contract to start the third year of work. I was tasked to conduct the audit and account for every penny spent during year-two.

The audit took about three weeks to complete and required me to view and document every receipt the company had concerning every travel action we took to deliver the MTES upgrade to the 21 MSTC locations. The process was not nice or peaceful due to the argumentative and less-than-cooperative nature of the company's owner. He was upset that I was the only person across the table from him during the audit and he had to produce every receipt I requested. He repeatedly asked me where certain other people were. I kept responding by saying I was representing the government team, was simply collecting the information, and would soon be joined by the people he mentioned. I hated lying to him but I saw it as the only way to gain his cooperation. I remained respectful and accommodating during the process and concealed my inner I-told-you-so feelings. The results of the audit showed there was no issue. I submitted my report with all supporting documentation. However, the Contracting Officer still refused to allow the third year of MTES work to start. Although my audit showed no funding issues, the Contracting Officer assigned a second person to conduct an audit of my results and ultimately continued to delay approving the third-year of work until the time expired. Before the time expired, the company's owner recanted his request for additional funds and agreed no funding issues existed—said he made a mistake. The Contracting Officer, however, still refused to allow the third year to begin. We were never able to start the third year of MTES work. All of our earlier work done to prepare for the future contract years was for naught.

The MTES devices and the salaries paid to develop and deliver them amounted to millions of wasted taxpayer dollars. The personnel at the MSTC locations were told they were not required to use the MTES devices to verify the Army's First Responders were being properly trained. They were told to send the systems to the Army's junkyard called the Defense Reutilization and Marketing Office (DRMO). Without the MTES, the MSTC Instructors returned to their antiquated training processes where accountability was questionable. On a lesser note, I no longer had a project to lead.

The MTES prime contractor was hit the hardest, I was told. The company no longer resided at the office where we held our meetings. When the government team members went to retrieve the government-owned lab MTES equipment from the company, according to a couple of my MSTC team members, they were given directions to the company owner's house. The government furnished property—the lab MTES devices—were stored in his garage. Although I felt the owner caused the situation by running to the "man", ruining his company was a price too high. I felt what my team did was borderline illegal. Years later, I searched the internet and was happy to see the company had been hired to perform under multiple other government contracts. I hoped the company was treated better on those contracts than my team treated them on the MTES contract.

I was disturbed to see a description of MTES, a couple of years later and with a new name, as a government need. Actually, the need surfaced in a document before the original MTES DRMO process completed. The new system was named MeTER—Medical Training Evaluation and Review System—and was included as a requirement on a different medical project called JETS. I wondered which company was going to win the new contract. Maybe the Puerto Rican company who served as a sub-contractor on the MTES project. Regardless of who won the new contract, the taxpayers would have to buy the system again.

The Spider And The Spout

Nettie was forced to return the project she gave to Sally back to me due to reduced productivity. Sally was moved to a smaller effort. Without MTES, I had plenty of time to complete the tasks on the project Nettie returned to me. I recognized the pattern of being placed on ailing projects, making the project healthy, and being removed from the project before the success is indisputably connected to my efforts. The climbing to the top only to be pushed back to the bottom to start again reminded me of an old nursery rhyme called the Eensy Weensy Spider. The spider repeatedly climbed up the spout only to be sent back to the bottom by the rain. After I completed the project Nettie returned me to, she decided to bite-the-bullet and completely remove me from the MSTC Lead Engineer position. Technically, she turned the clock back on the MSPB's order by returning the MSTC Lead Engineer position back to Sally. I was placed in the lower-level position Sally previously held. When I questioned Nettie about the switch, she simply told me I would no longer be the lead on the MSTC team. Nettie gave no reason for the switch. Although I filed a third complaint with the MSPB, I remained in the non-Lead Engineer position for years. I watched Nettie bring in additional Engineers to assist Sally with the tasks I used to complete by myself. Eventually, Sally had to be removed again for productivity reasons. She was placed back on the team where I served—her old team—and we both used our talents to view online catalogs and assist in locating and purchasing training materials for the Veterans Health Administration (VHA). It was work an eighth-grader could do.

The VHA team was led by a black man (Brock) who was sent to our division based on his past accolades (certifications, special training, and leadership positions held in other military organizations). Brock was sent to us with the expectation that he would serve as the Project Director. Instead, LTC Christ and his lackeys attempted to make Brock a Logistician with no leadership responsibilities. Since Brock had high-level friends including an outspoken black female Army Major, LTC Christ was warned that Brock would be sent back to his old organization in Alabama if he was not allowed to hold a Project Director position. Yes; the Major issued the warning to the Lieutenant Colonel. Our team leaders were forced to accept Brock as a Project Director. Brock and I were moved to an office (we were office-mates) when we relocated to a different building. I was able to play a small leadership role while working for Brock.

Brock made sure our VHA team was successful in acquiring and delivering training solutions to the VHA over multiple years. Brock represented the VHA team at conferences and other formal engagements. In preparation for a distinguished visitor, Brock perused his knowledge concerning all of our VHA projects. He expected to put on a good show for a Navy Admiral and her entourage who were scheduled to visit us the following week. On the day our visitors arrived, LTC Christ informed his full team (MSTC and VHA divisions) that we should proceed to the auditorium and be in position to greet the Admiral when she arrived. We all did as instructed and were in place and dressed to impress (no jeans or halter tops). Brock went a step further and wore a nice two-piece suit and tie. He offered several responses to the Admiral's questions and it was clear Brock was a leader among the team members. Actually, he introduced himself as the VHA team leader as we all took turns introducing ourselves to the Admiral. Near the end of the hours-long event, the Admiral thanked us for our hospitality and the great ideas we shared with her that might be useful for standing-up her new organization named the Defense Health Agency. Although we did not know at the time, the Admiral had two coins to give to the leaders of the MSTC and VHA teams. LTC Christ had given her the names of his leaders. Although Brock should have been summoned to join the Admiral at the front of the room to receive a coin on behalf of his VHA team, LTC Christ gave the Admiral the name of a white gentleman (Lenny). Lenny worked for Brock on our VHA team. Ironically, Brock had been complaining to LTC Christ that Lenny was not completing his tasks in a timely manner and was often away from the building when he should have been in his office completing assigned tasks. Not only was Lenny not the leader of the VHA team; his service in his non-leadership role proved to be problematic. I was convinced LTC Christ told the Admiral Lenny was the leader of the VHA team to upset Brock. Racial discrimination likely played a role in LTC Christ's action.

LTC Christ, in an effort to make it appear he was leading a large complex division, brought in additional team members and placed them in leadership roles. The new team members had a history of mistreating blacks in the division from which they came. One of the new members was Judas. Judas was brought in to serve as LTC Christ's Deputy. Judas brought in one of his contracted employees (Seth) from his old division. Seth was hired as a new government employee and was placed on the VHA team. Within months, Seth was given several of Brock's projects and deemed a second leader of the VHA team. I was assigned to Seth as his Engineer.

Seth retired from the Army as a Lieutenant Colonel less than a year before being hired by my organization as a government contractor and then a government civilian. The rumor was Seth applied for the government position but scored amongst the lowest qualifiers. Seth was highly favored by Judas and would stroll past my and Brock's door each morning on his way to Judas' office for a chat. For about a month, Seth would blurt out "What's up dudes in da hood?" to me and Brock as he passed our office door. After a few times, I told Brock I was going to say something the next time he greeted us in that way. I felt he viewed us as the characters in the movie Boyz N The Hood. I told Brock I believed, regardless of our accomplishments that ran counter to the movie's black male characters, Seth could only see us as the nefarious characters who often broke the law and had no aspirations beyond the available aspirations within their neighborhood. Brock decided he would not give Seth another opportunity to degrade us. He asked Seth if they could have a word in the vacant office down the hall. Brock was far more outspoken than I was. I could hear the sounds coming from the office but I could not understand what was being said. I wondered if I should enter the room to break the flow of the somewhat loud conversation to ensure there was no escalation. The result of the conversation was Seth never referred to us as "dudes in da hood" again. Instead, he referred to us as boys and girls. Especially if a female was in our office. Seth obviously had a serious issue with viewing and respecting me and Brock as men—and as his equal.

Serving as Seth's Engineer was not a pleasurable experience. Seth refused to accept my input on technical issues and almost always changed the wording in my technical documents. In many cases, his changes showed his gross lack of understanding concerning basic acquisition rules. One day he asked me to explain why we needed separate documents for statements of work and system specifications. I explained the differences to him. I told Seth the statement of work shows what is required of the defense contractor (e.g., monthly meetings, schedule deliveries, etc.) to meet the requirements of the contract and the system specification shows what is required of the training simulator (e.g., computer speed, vehicle simulated, etc.). Seth still believed the documents should be combined and decided to seek an answer to the question from other Engineers. I listened to another Engineer tell him the same thing I told him.

One VHA project to which Seth and I were assigned entailed developing and delivering a patient-interviewing training simulator to be used to train VA medical professionals in the art of interviewing patients. I worked that project with Brock for a year before Seth was hired and the project was taken away from Brock and given to Seth. Seth repeatedly changed my documents even though my words were correct and traceable back to agreements made with our customer before Seth was hired. I would try to explain the history of my statement of work document and system requirements document entries to Seth but he would disagree without sound reasoning and go to my Supervisor, Nettie, for support. Of course, Nettie never took my side in the disagreements. Although Nettie approved the documents before Seth was hired, she would either make attempts to rationalize Seth's position or simply tell me to work it out with Seth. The results always meant Seth would get what Seth wanted. I viewed Seth as a one-man-band. He was the type of leader who had to play every position on the team. Eventually, Seth cut me out of the process completely and completed all of the engineering and management documents required to award the contract. Our contracting personnel used Seth's faulty documents and awarded a contract that was problematic from the start, had to be modified several times to correct Seth's mistakes, and resulted in the VHA not getting what they paid us to acquire for them. Ultimately, the taxpayers paid for a product that was not useful. Regardless, our team was given a Team of the Quarter (TOQ) award for awarding the contract that led to the less-than-useful patient-interviewing training tool.

Seth nominated our team for the TOQ award. When he listed his team members, he purposely left me off the list. Seth and our division's chain of command, others who approved of my name being omitted from the list of team members, knew I had worked the project for several years and had revealed major issues with the contract that resulted in badly needed corrections. Even still, they all allowed my name to be omitted from the team nomination list. I was not aware our team had been nominated until I saw them seated in the TOQ section of the auditorium during one of our organizational "town hall" meetings. When my team won, was congratulated on stage in front of the entire workforce, and given t-shirts and other prizes, I was angry and hurt. Soon after, I asked our division's leader (COL Scotty) to correct the mistake but he ignored me. COL Scotty had recently replaced LTC Christ as our leader. When I mentioned the TOQ issue to Judas, now COL Scotty's Deputy, he told me our organization did not have enough t-shirts for all of the team members so I was left off the list. This proved to be another one of Judas' lies. I complained to higher-level organizational leaders and they corrected the issue. Seth was eventually moved to larger projects in our medical simulation division. I told Brock we were looking at our future leader. It was clear to me Seth had been placed on the fast-track to upper management. It did not matter that it was obvious Seth's experience and mental acuity levels were lower than ours; we were viewed as being inferior and treated as such.

I realized people do not care if you are aware of their shortcomings if they view you as inferior (i.e., believe your issues exceed theirs). Actually, I witnessed this while working for the Navy's Training and Simulation organization. One of my very attractive coworkers came into my cube and confessed that she was cheating on her husband with another employee. She started by telling me I probably had done worse. I was anxious to hear her next words and didn't dare tell her I had not done many bad things in my life. I listened to her purge her soul of her wrongdoing and watched her cry. After she finished speaking, I advised her to stop seeing the guy and reminded her of the importance of maintaining her family atmosphere for her young son. I advised her to stay with her husband. I added that she should never tell her husband about the affair. For weeks I was her shoulder—even when I didn't want to hear about her continued adultery. It was clear to me she could not stop seeing the guy. I began to feel like a Priest listening to her work-interrupting confessions. In fact, she told me she solicited help from her church but when the church's Counselor said she would have to start paying for the sessions, she stopped meeting with him. The only way I was able to get my very attractive coworker to stop

confessing to me was to show interest in her myself. Turned out I wasn't her type.

Several of my Army civilian coworkers began to open up to me about their issues with our

organization's leaders. Once the word got around that I filed several complaints against the

organization, many of my coworkers who were suffering at the hands of our leaders felt

comfortable sharing their stories with me. I heard stories from people who I thought, based on

their demeanor around the organization, were happy with their working conditions. One lady told

me she was sexually harassed in the parking lot. Another told me an Army Major exposed his

private parts to her on the elevator. I was also told about a Colonel who touched his female

employees inappropriately and took under-the-table photos of one female who was wearing a

skirt. That Colonel was placed on "special assignment" and allowed to retire. The woman whom

he took photos of received a promotion to a higher-level position. Another coworker reported

that numerous employees were not working the 8-hours-a-day shift we were required to work

and were fraudulently claiming the 8-hours-a-day on their timecard. Many of us knew of at least

one person who was cheating on her timecard. The timecard fraud allegation led to an Army

investigation, known as an Army Regulation 15-6 (AR 15-6) investigation, that showed roughly

80 people were found to have cheated on their timecard. AR 15-6 investigations are used to

ascertain facts and report them to the appropriate appointing authority. My organization's leaders

countered the results of the AR 15-6 investigation by saying the badge-scanning equipment was

faulty and decided to only punish one person for misrepresenting his work time. Years have

passed and we are still using that "faulty" badge-scanning equipment. The stories my coworkers

told me left me in shock. I could not fathom why my organization's leaders collaborated on

punishing/harming me without having any proof that I violated any law, rule, or regulation but supported other employees where the results of an investigation showed they violated laws, rules, and/or regulations related to sexual harassment, sexual assault, or fraud. I felt something should be done about the behavior (i.e., misbehavior) of our organization's leaders.

I decided, as a taxpayer, to send a request for assistance to four (4) congress members from Florida (Stephanie Murphy, Darren Soto, Bill Nelson, and Marco Rubio). My letter started off by congratulating the congress members on passing Bill S. 585 allowing a 3-day suspension of Supervisors who retaliate against employees. I ended my letter by saying, "Employees at my organization have suffered retaliation for reporting unsubstantiated low performance review scores, demotions with no supporting business reason, unfair hiring practices, time card fraud, and even sexual harassment." I included my address and telephone number and informed each congressperson he/she could contact me if they wished to discuss the contents of the letter. Only Senator Nelson responded—via a letter. He thanked me for the congratulations but did not mention the issues I raised in my letter. Although I received nothing from the other three congress members, a coworker told me she was asked by one of representative Murphy's assistants if she knew me. At the time, my coworker was working with Murphy's office on another matter.

I applied for a promotion at my organization where, if promoted, I would have been required to supervise other Engineers. Since I had 26 years of experience, my master's degree was in management, and my doctorate degree was in organizational leadership, I felt I was definitely qualified for the position. I looked forward to putting my leadership knowledge to work in a place where good leadership, as defined by my doctoral studies, was virtually non-existent. I was happy when I received an email message saying my resume had been forwarded to the selecting official as highly qualified and I would be considered for an interview. A few months later I received a second message saying I was not chosen for either of the positions—one position became two. I asked for a debrief in response to the second email message so I could find out why I was not selected for the position.

I was given a debrief by the leader of the selection panel. He told me my resume did not score high enough to meet the level designated as the cutoff for qualifying for an interview. The information was shocking to me. In his explanation of why I was not selected, the selection panel leader told me the winner had a doctorate degree and had published papers in the past. The winner had also presented his papers at several conferences. Although he named the two white persons selected for the positions, the selection panel leader never mentioned the qualifications of the second selectee. Since I knew holding a doctorate and presenting papers at conferences were not required to win one of the positions, I mentally questioned why I was not selected or allowed to interview for the positions. Also, I was aware the person who was granted the second position did not have a doctorate degree. I later found out she did not have a master's degree either. I thanked the gentleman for the debrief, returned to my desk, and filed an EEOC discrimination and retaliation complaint—my second EEO complaint. I believed I was not selected to interview for the positions because of my past EEOC and MSPB complaints. I later found out the selection panel leader created a new requirement after receiving my resume and found that my resume did not meet his new requirement. Therefore, he never allowed my resume to be formally reviewed—I was not allowed to compete for the promotion position(s). The new requirement directly opposed a requirement that was included in the job announcement I applied against. My resume met all of the requirements included in the job announcement.

The selection panel leader's Supervisor served as the Selecting Official (i.e., the person who selects the winner of the competition). The Selecting Official was Mr. Skippy—formally LTC

Skippy. Mr. Skippy was actually the backup Selecting Official. He was told to serve as the Selecting Official after his Superior, A Colonel, became embroiled in the photos-under-the-table sexual harassment complaint I mentioned earlier. With Mr. Skippy as the Selecting Official, I am not surprised my resume was unfairly omitted from the competition.

LTC Skippy and I had past issues and he undoubtedly had friends in higher-ranking positions in the organization. I say undoubtedly because LTC Skippy, reportedly, had his high-level civilian position waiting for him when he retired from the Army. A prospective mentee of Mr. Skippy told me Mr. Skippy, after being given his high-level position, was chosen to mentor lower-level managers to facilitate their climb to higher-level positions such as the one given to Mr. Skippy. According to the prospective mentee, Mr. Skippy told a group of prospective mentees he was hired at the highest level and probably was not the best person to tell lower-level employees what they needed to do to get to his level. Mr. Skippy, in my opinion, was also not the best person to serve as a selection official for positions where the official was expected to recognize applicants who legitimately worked hard climbing their way through the ranks in anticipation of being rewarded via a well-deserved promotion. Because of the unlawful actions of Mr. Skippy and his associates, I was forced to continue serving on the low-level VHA projects.

Serving on the VHA projects gave me a clear understanding of why several investigations were employed to uncover questionable activities at the Veterans Administration (VA). The inept and unethical leadership at the VHA's training division, a subsidiary of the VA, led to several violations related to acquiring equipment and material to be used for training purposes. Two of the high-level VHA training division leaders were hired after they served as leaders at either the Army training division where I worked or the Navy training division where I previously worked. They were both asked to leave the VHA following investigations concerning possible criminal activity. My organization eventually dissolved the partnership we had with the VHA and, reportedly, returned all remaining funding. I often wondered if my leaders feared the reported "witch hunt" at the VA would reveal criminal activities at my organization. My suspicions were likely correct since my team leaders secretly maintained some of the VHA funds and used them to pay me and some of my coworkers for several months after the partnership ended—even after we were moved to non-VHA projects. Apparently, my leaders were fearless.

We Need A Union

There was growing talk concerning the need for a union at my organization. Numerous employees wanted and needed help as a result of several organizational leaders acting outside of the best interest of the taxpayers and the organization's employees. Because several of the employees approached me about starting a union, I decided to query all of my coworkers to gauge union interest. In the past, I served as a Steward in the AFGE union at the Navy's Training and Simulation organization. I expected to gain enough signatures to start a local union at my current organization. I sent an email message, from my work computer at lunchtime, asking all of my coworkers to respond so I could get a count of how many of them were interested. The responses started coming in immediately—mostly in favor of a union. I also received responses against starting a union. Some of the negative responses against starting a union included curse words. My organization's leaders, in response to my message, suspended my computer access to the Army's network and left me unable to complete my work duties for that day. I sat around until I completed my eight-hour shift and then went home.

My computer access was restored by the time I returned to work and my Supervisor issued me a letter stating they decided I had not violated any rules by sending the initial union-interest email message. The letter also informed me that I should remove anyone who asks to be removed from my union email list. Because my leaders appeared to be aware that several people requested that I remove them from my union email list, I wondered if they had intercepted any of the responses to my initial message. I decided to send a second message to my coworkers, from my personal email account, informing them that my work email account had been disabled for a while and I was not sure if I received all of the responses they sent to my work account. I asked that they resend, to my personal account, any responses they sent to my work email address. My leaders found out about my second message and became angry concerning my insinuation that they might have blocked some of the union-interest responses addressed to my work account.

I was able to gain assistance from a coworker in contacting the American Federation of Government Employees (AFGE) union district office. The coworker arranged a lunch meeting with an AFGE representative to discuss our next steps. The representative explained the process to us and said she would hold a meeting to explain the process to the larger group of employees who were interested in starting a union. I sent a message to my organization's employees inviting them to the AFGE meeting. However, on the day of the meeting, only a handful of my coworkers showed up. I expected to at least have the 20 to 30 people, who told me they would support a union, walk through the door. But that did not happen. I guess pizza and popcorn were not strong motivators. Although I used the signup sheets the AFGE representative left with me to continue trying to get enough signatures to start a union, we were unsuccessful. The accusations claiming my organizational leaders had interfered with the union-starting efforts must have been true. The effect was catastrophic. In response, I filed a complaint with the Federal Labor Relations Authority (FLRA) alleging my leaders violated its employees' right to attempt to start a union without interference from the agency.

The FLRA governs labor relations between the federal government and its employees. The FLRA adjudicates disputes and decides cases concerning allegations of unfair labor practices leveled against federal agencies by federal employee unions. Because our current President (Donald Trump) has not appointed a General Counsel to the FLRA, the FLRA cannot process unfair labor practice complaints. The lack of a General Counsel means unfair labor practices cannot be investigated and the perpetrators cannot be prosecuted when warranted. President Trump indicated he is in no hurry to fill the General Counsel position. To date, the FLRA has been without a General Counsel for over two years. Therefore, thousands of federal employees, including the ones in my organization, have been left unprotected and potentially in an unsafe working environment.

I did receive, in response to my initial union-interest message, several requests to be removed from my list. I was sure I removed everyone who asked to be removed. It turned out I had not. I later found that I mistakenly left one person (Hawkeye), who asked to be removed, on the list. When Hawkeye received a second message from me, she responded by reminding me she asked to be removed from my list. She then sent a complaint to my Supervisor. At the time, I was not aware Hawkeye worked directly for my organization's leaders. I had not yet considered she was knowingly used to assist my Supervisors in their next attack.

Nettie forwarded Hawkeye's message to me and ordered me to remove Hawkeye from my list and send Hawkeye a message informing her she had been removed. Because I held very little trust in my leaders due to their past actions—including suspending my network access, I told my Supervisor I was not comfortable sending Hawkeye the union-related message from my work computer. I told her I feared retaliation. Nettie responded, including her Supervisor (Judas) on the email message, by telling me she was offended that I would think she would retaliate against me. I was surprised by her response since we both were aware I had an ongoing retaliation complaint filed against her with the MSPB. Judas sent me a follow-on response telling me Nettie would only retaliate against me if I did not follow her order. Judas' response caused me to lower my intellectual view of him. I was surprised he admitted my leaders would retaliate against me. I gave him additional points, though, for honesty.

Nettie decided to issue me a letter reprimanding me for not removing Hawkeye from my list and for not sending Hawkeye a message saying I removed her. Not once had she asked me if I removed Hawkeye from my list. She had a witness in the room when she issued me the letter. I was not smart enough to bring a witness with me. We had a short discussion concerning her letter. During the discussion, I told her I had removed Hawkeye but wished to speak with my Attorney concerning sending the union message to Hawkeye as she directed. I asked Nettie to rescind the letter since it incorrectly accused me of not removing Hawkeye from my email list. Nettie refused. After the meeting, I sent Nettie an email message covering the discussion we had and again asked her to rescind the letter. Nettie responded with a formal letter saying she refuse to rescind the reprimanding letter she issued me. Later that same day, three of my coworkers told me they could hear the meeting conversation I had with Nettie through our thin wall. Our offices were actually separated by wall-like room dividers. I wasn't aware I would, later, need those coworkers to recollect what they heard.

I sent a message to Judas asking him to rescind the letter Nettie issued me. In my message, I outlined several issues concerning Nettie's letter. The letter threatened severe formal disciplinary actions against me if I took a similar action in the future. Also, Nettie told me in writing she planned to keep the letter in her files forever. Judas decided against rescinding Nettie's letter. Once Judas and Nettie found out I chose to share Nettie's reprimanding letter with several of my coworkers, Judas issued me a stronger-worded reprimand and had it placed in my personnel file for two years. Judas' letter made an additional claim that I threatened Nettie during the three-person meeting wherein Nettie issued me her letter with her witness in attendance. Judas' letter also stated Nettie's letter was based on her assumption that I had not removed Hawkeye from my email list. Judas' letter was full of fabrications. Shortly after Judas issued his reprimand letter to me, Nettie used both reprimanding letters to justify giving me the lowest annual performance review I ever received in my DOD career. I remember the enlightening thought that the Army's Training and Simulation organization had become the worse place I ever worked. I was forced to file a third retaliation complaint with the MSPB.

The union-related reprimand MSPB case progressed to the evidentiary discovery phase. During the discovery phase, my organization was forced to give me numerous email messages exchanged between Judas, Nettie, higher-level leaders in my organization, and some of my coworkers. The letters showed the collusive actions taken that led to the reprimanding actions of Nettie and Judas. Their conspiring efforts were deemed successful in Judas' final email message to his conspirators. Judas informed them that he met with me to deliver the reprimand, he read the reprimand letter to me, I had nothing to say in response, and I refused to sign the reprimand letter. He added that he also counseled the Secretary he believed was helping me. He ended the email message by saying, "Thanks everyone for your help in crafting these letters. Great staff work."

Unlawful Disclosure

Throughout all of the negative actions my "leaders" took against me, I continued to perform successfully on my projects. I believe my near-impeccable performance was driven by my quest to show my leaders were lying about me and to force them to create new lies in their continual efforts to keep me from progressing professionally. Several people questioned why I continued to push hard in my duties. They knew I was being mistreated. I often responded by telling them "I work for the taxpayers and Soldiers." I was able to keep my feelings concerning my leaders separated from my feelings concerning what I needed to do to successfully meet the Army's mission. Occasionally, people would acknowledge my mistreatment and then share useful information with me. They would tell me when certain secret meetings were being held about me or name coworkers who were also being mistreated by our leaders.

One day a coworker asked to meet with me in confidence. I did not know him so I was skeptical about meeting with the gentleman in an out-of-the-way place. We met in the first-floor break room that was seldom used. When we met, he told me about some of the issues he was having. As usual, I listened and empathized with his situation. I was acutely aware of the extent our leaders would go to harm an employee who was not in their inner-circle. We reached a point in the conversation where my coworker opened the folder full of documents he was carrying and pulled out a multi-page letter that looked familiar. He told me he had read my formal EEO complaint (the first one I filed concerning the Bradley team). When he handed me the document, I scanned it and realized it was my formal unredacted complaint. I was shocked! My first thought was how in the hell did he get my complaint? Before I could ask the question, the gentleman told me he printed my complaint from our organization's common drive that is open to all employees in the organization—including the contracted workers. He also told me who uploaded my complaint to the drive. I sat quietly not knowing how to respond. My thought was everyone knows about my allegations against certain organizational leaders who might be loved by many of my coworkers. They also knew about my admitted mental distress and other sensitive/confidential information such as my social security number, home address, telephone numbers, etc. The complaint must have been released to a coworker by the organization's Lawyers. I thanked the gentleman for sharing the information with me and agreed to answer any procedural questions he had in the future concerning filing grievances or complaints against the people who were mistreating him. Once I returned to my desk, I searched for and located my complaint, along with several other legal documents in a folder labeled "Harroll Ingram Litigation", on my organization's common drive. I immediately reported the security infraction to our EEO office representative. The legal folder containing my unredacted complaint was still on the common drive a week later. The lack of responsiveness led me to file a third EEO complaint. The folder was deleted after I filed my EEO complaint and nine days after I asked to have it removed.

I began to wonder how many people read my complaint. I also wondered if those who read it got angry at me for the words I included in my complaint and if they reacted negatively toward me in retaliation. In my mind, that explained why people who used to say hello to me simply looked the other way when I approached. It also might have explained why my resume was unjustifiably thrown-out of a past competition for a promotion position. One of the selection panel members did testify under oath at the EEO fact-finding conference that he knew about my past EEO complaint. I also wondered about the three times a strange vehicle stopped in front of my house, the time I pulled up to my house and found a strange vehicle backing out of my driveway, and the night a Police Officer knocked on my door and told me a neighbor reported a strange person looking through my window. Many thoughts ran through my mind. However, I had no way of knowing if a relationship existed between those who read my complaint and the trespass actions at my home. My mind reverted back to what my elders and experience had taught me; a black man rising in predominately white settings is usually forced to deal with racism and might be in danger. I was undoubtedly in a predominately white setting at work. The statistical hiring evidence collected during my initial EEO case, the one that was posted for everyone to see, showed that less than 4% of the Engineers at my organization were black. Only one black Engineer had been promoted above my level in the past few decades. It was no wonder why my promotion path was being blocked.

Unlawful Hiring Practices

I sought to have the EEOC investigate the disparate and discriminatory hiring practices at my organization but they refused. one Judge simply ignored my request (i.e., ignored that part of my complaint). A Judge in one of my later cases decided he would not address the claim because, according to him, I failed to include supporting statistical data in my legal submissions. The truth was I had included statistical information and referred to the information in my legal narrative. I included documents showing the nationalities of the Engineers at my organization. The organizational demographics documents were submitted to my legal team by my organization's Lawyers during the discovery process. My narrative highlighted that some of the Engineers listed in the document were coded as African American but were actually Asian or Latino. A separate statistical document from my organization included a statement at the top informing that white female Engineers were omitted from the list. Omitting the white females strategically gave the appearance there were fewer white Engineers than there actually were and skewed the comparison statistical data. Even still, the number of black Engineers compared to white Engineers were discriminatorily low. The Judges chose to ignore that I had submitted the statistical evidence.

Circumventing fair and legal hiring and promoting processes is not as difficult as one might think. I have seen and been told of people being hired at a higher level than their qualifications justified. Some job announcements were written to fit a particular person's qualifications. There have also been cases where the targeted person was coached on how to pass the interview—even given the interview questions in advance. Selecting Officials have gone so far as to post the job as if it was in a different state to ensure local would-be competitors would not apply for the position. To make that tactic work, the desired applicant was told how to locate the job announcement and told to apply. As the only applicant from the local workplace area, the desired applicant got the job. How had my organization maintained the lower than 4% hiring rate for black Engineers? Easily, I suspect.

One black gentleman who worked in my organization's contracts division applied for a promotion and was selected as one of the two qualified finalists. The other finalist, a white gentleman, was chosen for and offered the position. However, the gentleman declined the position. Rather than offer the position to the black guy, the sole remaining highly-qualified person, the Selecting Official issued a new job announcement seeking more applicants. When the same black gentleman was again selected as one of the two qualified finalists, the other finalist, a different white gentleman, was chosen for, offered, and accepted the position. The black gentleman filed an EEO complaint.

A black logistics employee who worked for the Navy's Training and Simulation organization, a stone's throw from my organization, applied for a leadership position in her organization and was chosen as one of the applicants to be interviewed. She completed the interview process but was not chosen for the position. One of the other finalists, a white woman who also failed to get the position, told the black woman that all of the white finalists, including the winner, were given a "mock" interview before the bonafide interview process started. They were even allowed to hear the actual interview questions beforehand and practice their responses. If these unfair tactics exemplify recruiting processes around the country—not just in Central Florida, there is little wonder why blacks are sparsely represented at leadership levels in the federal government. The selection process is often faulty. In some cases, the process is heavily influenced by a concept known as groupthink.

Groupthink occurs within a group of like-minded people where decisions are made based on the group's common desires. Because of the likeness of the people chosen to serve as part of the group, alternative views are usually not seriously discussed. Alternate views are actually suppressed when groupthink is at play. The outcome is things remain the same. When groupthink invades an employment position selection panel, the position is usually filled with a person who thinks like the people serving on the panel. The people hired and promoted behave and think like the existing people in the organization. Groupthink is very harmful to creativity, uniqueness, and, at my organization, diversity.

Healthy Living

I began to focus on my health in 1990 when I stopped eating red meat. I only ate fish and fowl during that year before adopting a full vegetarian lifestyle in 1991. I believed becoming a vegetarian would allow me to live a long pain-free life and I would escape the ailments I often saw in my family such as high blood pressure, diabetes, cancer, etc. Of course, I was wrong. As I aged, I realized I had the same painful joints meat-eaters had. I eventually experienced short periods of high blood pressure and borderline high blood sugar levels while working for the Army's Training and Simulation division and defending myself against the unlawful actions my leaders took against me. Why was this happening to me? I believe the answer was I failed to factor-in stress. My stress levels were ultra-high as a result of being subjected to discrimination, retaliation, and general (as well as professional) disrespect at my organization. I experienced difficulty sleeping at night. On some nights, I had to sleep with a trash can by my bed because of nauseous feelings. In 2018, my right shoulder starting hurting for no apparent reason and continued to bother me for over a year. My doctor noted my issues, prescribed me sleeping pills, and had her assistant take an Electrocardiogram (EKG) reading to make sure I had not suffered a heart attack. Blessedly, the EKG result did not show a heart attack—according to the assistant. One question I was often asked was why didn't I leave the Army's Training and Simulation division. I always gave an answer but rarely the one that immediately popped in my head. The real response was why should I have to leave. I was a government employee paid with tax dollars collected from citizens of my country. My ancestors paid some of those taxes used to pay government employees' salaries. The problem resided with the people who chose to assist those who were repeatedly attacking me and causing my stress levels to increase which facilitated my unhealthy state. The question of leaving should have been directed at those people rather than me. I chose to fight back rather than cower in the face of racism and bullying. The great orator and writer, Mr. James Baldwin, wrote in his book titled Confronting History, "Not everything that is faced can be changed, but nothing can be changed until it is faced." I faced the racist cats in an effort to facilitate change.

Facilitating Change

Earlier in my career, while working for the Navy's Training and Simulation organization, I told the Head of Engineering about the issues I was having with my Superiors (his subordinates) and asked him for advice on righting things. I was surprised to hear his one-word response. He replied, MOVE! That was it. While working for the Army's Training and Simulation division, one of my mother's friends gave me similar advice. We were at my aunt's wake when she told me my mother told her about the unfair treatment I was being subjected to by my leaders at the

Army's Training and Simulation organization. Having worked as a high-level government employee in Washington DC, she was aware of the various tactics used to get rid of unwanted employees. My mother's friend told me, "You should leave. They have nothing else to offer you and you have nothing else to offer them." I often wondered if I should have heeded the advice.

My health began to improve as I took control of my life and changed how I physically, mentally, and emotionally responded to the continued attacks. Once I accepted that my attackers feared the elevation of black men, were just reacting as they were taught to react by their elders, and refused to recognize that not all black men took back-lashes with a smile, I was able to understand the need to use the emotional intelligence skills I was taught during my doctoral studies.

Emotional intelligence makes use of five key elements that assist in adapting to environments. Those elements are Self Awareness, Self Regulation, Motivation, Empathy, and Social Skills. Self Awareness helped me understand my emotions, values, and goals. I began to follow my inner feelings (often referred to as guts) in response to being attacked. Self Regulation helped me to properly direct my responses in ways that produced short-term and long-term value. Motivation pushed me to do what I knew was right even if it led to periodic short-term stresses. Empathy ensured I did not over-exert in my responses to my attackers. Lastly, Social Skills pushed me to renew and create relationships to facilitate providing assistance to anyone who needed it as well as seeking assistance from others in reaching some of my goals.

The greatest credit I can give for taking control of my life and warding-off the unhealthy stresses of my work environment goes to my family. My immediate family watched closely as I fought to defend myself against the unwarranted attacks (attacks I often say resulted from uncontrolled racist urges) perpetrated by my coworkers and leaders. In some instances, my family asked me to "let it go" in hopes that the perpetrators would end the attacks. However, after witnessing the continued attacks as well as the evidence showing I had not done anything to deserve being attacked, my family accepted that I had to defend myself. They offered emotional and spiritual assistance throughout the processes. My little doggie, a Maltese-Poodle named JJ—short for Juney Jr, was with me every step of the way. He watched me prepare my MSPB and EEO cases on my laptop and rode with me to the post offices to mail my legal submissions to the adjudicators. JJ's head was the first thing I saw each day I returned home from battle (i.e., work) and opened the front door. The sight always shifted my focus to positive thoughts. I could swear he knew I was greatly bothered and continuously offered closeness to successfully reduce my stress. I am not sure I would have made it through the storm successfully without my family.

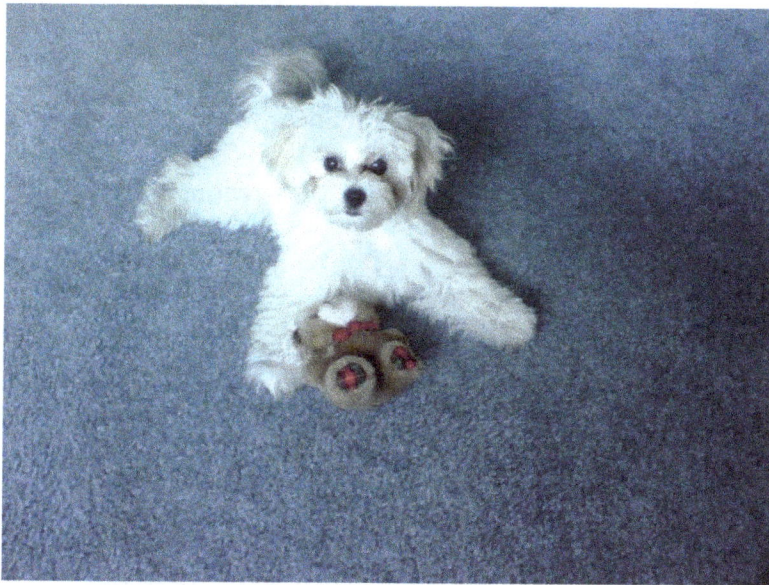

Where I Am

I live in Florida where appeals of MSPB and EEOC decisions must be filed in a particular circuit court. I filed appeals concerning multiple decisions with the U.S. Court of Appeals for the Federal Circuit. My appeals were either disposed of via a tactic known as a Summary Judgement or I was issued an unfavorable decision by the Federal Circuit Judge(s). Reportedly, the U.S. Court of Appeals for the Federal Circuit has a history of issuing unlawful/inaccurate decisions.

The Federal Circuit was created by congress in 1982 to address the numerous issues related to poor judicial decisions concerning patent law. The circuit court was not initially intended to hear appeals such as those related to merit systems or equal opportunity. Subsequent congressional debates resulted in a decision to include additional jurisdictions (i.e., authority to hear non-patent cases) on the Federal Circuit and view that court as a general court—not specialized. Relegating the Federal Circuit to one comprised of Judges who do not have a specialty could have facilitated the issuing of less-than-lawful decisions due to misunderstandings in certain legal areas.

In an article titled Supreme Court Reversal Rates: Evaluating the Federal Courts of Appeals, the Federal Circuit Court was given a "D" grade due to the number of decisions reversed by the U.S. Supreme Court. The Federal Circuit Court's performance was viewed as the worst when compared to all the other Circuit Courts. Because the only recourse for people who received an unfair ruling from the Federal Circuit Court is the U.S. Supreme Court and many people/Lawyers do not find appealing to the U.S. Supreme Court to be advantageous, many of the less-than-lawful decisions made by the Federal Circuit Court stand, resulting in a revictimization of the people who appealed to the Federal Circuit Court. Not all of my appeals were filed with the Federal Circuit Court.

I appealed my first EEO case to the U.S. Court of Appeals for the Eleventh Circuit. The Eleventh Circuit Federal Court has jurisdiction over federal cases in the middle district of Florida where I reside. Multiple Lawyers told me the Judges in the eleventh circuit were generally not fair. One told me only three were fair. He added that he would take my case if I was assigned one of the three fair Judges. Another Lawyer told me of a colleague, another Attorney, who moved north to practice as a result of too many unfair and unlawful rulings from eleventh circuit Judges. A third Attorney warned me that I would likely not get a just decision in the eleventh circuit, especially while Donald Trump was in office. Despite the warnings, I felt the need to continue with my appeals. The evidence clearly showed unlawful actions were taken against me by my leadership. I was sure there was no way the federal appeal courts could rule against me. I was wrong.

Complaints Resolution

I was forced to seek assistance in defending myself against my organizational leaders over a 12-year period. I requested help from multiple government organizations in response to illegal, unethical, and unhealthy conditions in my organization. I filed a total of eight (8) formal complaints; four (4) with the MSPB, three (3) with the EEOC, and one (1) with the FLRA. Although the MSPB and EEOC Judges' responses can be easily viewed via any online search engine (e.g., google, yahoo, etc.), my filings were not made available to the public. In most, if not all, cases my filings show the MSPB's and EEOC's final responses were at least highly questionable—unlawful in my view. The MSPB's and EEOC's responses left my leaders feeling they were above-the-law and led to repeated actions of disrespect and unlawful behavior towards me. I included several of my submitted complaint process documents in the appendices of this book. The theme I recognized as a result of my complaints was federal organizations protect other federal organizations.

I lost (i.e., received an unfavorable ruling concerning) most of my complaints. It turned out, unfavorable rulings from the MSPB were not uncommon. Every year the MSPB reports statistics concerning the results of all of the complaints filed for the year. Each year, the percentage of favorable rulings (i.e., rulings in favor of the complainant) is revealed. The result is always below 10%. The Judges in my cases not only searched for any technicality to rule in favor of my organization; the Judges ignored evidence and ruled as if I had not submitted the evidence they chose to ignore. Those Judges showed no concern for the unlawful treatment I was subjected to by my leaders. The actions of those Judges reminded me of the Dred Scott v. Sandford decision of 1857 where the Supreme Court ruled, "The black man has no rights which the white man was bound to respect". Numerous other cases exemplified the same mindset. I watched the movie "How They See Us" last night and dealt with feeling of anger and sorrow. The movie revealed another example of how the "justice" system ignores the constitutional and human rights of black men in an effort to degrade and unlawfully punish them (us). I gained a better understanding of the depths of the hatred held by and the coldness in the hearts of those who consciously and unconsciously work to oppress black people.

The Equal Protection Clause of the 14th amendment of the constitution is supposed to guarantee equal protection under the laws of the United States. Although I was supposed to receive equal protection under the laws of the United States, my leaders and coworkers, as well as the MSPB and EEOC Judges, worked hard to circumvent those protections. I suspect many other whistleblowers have been re-victimized by MSPB judicial rulings. In many cases, the Judges' final orders, which are posted on the internet, do not address key evidence submitted by the complainant. So, the appearance is the complainant left the Judge with no choice but to issue the unfavorable ruling. That is often far from the truth. My cases support the notion that many Judges refuse to issue impartial rulings. I believe my skin- tone played a role in the unfavorable decisions issued by the Judges concerning my cases. Perhaps a black man presenting legal arguments to the court as a non-lawyer is viewed as not worthy of equal protection under the law. The black man's voice goes unheard before the court. Decades ago, women had the same issue in this country.

Supreme Court Justice Ruth Bader Ginsburg explained that her notoriety, earned in the early 1970s, was possible because she had a voice before the courts—as a Lawyer. She indicated she would not have had a voice as a non-Lawyer. Justice Ginsburg, known for fighting for gender equality, stated, "For the first time in history, it became possible to urge before courts, successfully, that equal justice under law requires all arms of government to regard women as persons equal in stature to men." I believe the argument can be made that Justice Ginsburg's statement applies to black men as well. Equal justice under law requires all arms of government to regard black men as persons equal in stature to white men.

It is possible that some of the unlawful unfavorable rulings result from inadequately trained Judges. It is important to note that some of the MSPB Judges do not have law degrees and have not passed the bar exam in any state. Licensed Lawyers have fought, unsuccessfully, to force the MSPB to require a law degree for their judicial positions. On the other hand, maybe the Judges are adequately trained and their inadequacy lies somewhere else. Former U.S. Supreme Court Chief Justice William Rehnquist might agree if he were still alive. He described one of his colleagues by saying he "was one of those people who are intelligent and learned, but seriously lacking in judicial temperament."

I filed my first complaint with the MSPB in 2008 in response to my leader on the MSTC team retaliating against me because I decided I would not escort a company to one of our MSTC locations so they could take videos of another company's technology to aid in the development of a future training solution. My organization's Lawyers explicitly forbade me from escorting the company to the MSTC for the planned purpose. The Lawyers informed the planned action could violate trade secret laws.

My leaders, in retaliation, issued me a low annual performance review score and removed me from my leadership position on the team. After viewing all of the evidence, including the petition document I submitted—see Appendix A, the MSPB ordered my leaders to increase my performance review score and return me to my leadership position on the MSTC team. The favorable ruling I received concerning my first complaint proved to be an anomaly. In fact, the MSPB Judge changed his ruling at the behest of my organization's leaders. My leaders refused to increase my annual performance review to the highest level as ordered. They chose to only increase it to a level below the highest level. My leaders refused to heed the Judge's warnings that he would have the refusing leader's pay withheld if his original order was not followed. Eventually, the Judge allowed my leaders to submit additional evidence after the case was over and subsequently ruled the level to which my leaders chose to raise my score was sufficient. I was, however, returned to my leadership position on the MSTC team. By that time, I had already filed my second complaint based on retaliatory actions taken against me on the team to which I was moved after I was removed from the MSTC team.

I filed my second complaint with the EEOC in 2011 in response to my leaders removing me from my leadership positions on the Bradley Gunnery Training Simulator team because I complained to them about a government support contractor who was usurping my government responsibilities, undermining my authority by secretly telling our customers I was lying to them about the status of their simulators, and taking other actions that fostered a hostile work environment. My complaint also informed of the hiring disparity concerning black Engineers at my organization.

The EEOC examined the submitted evidence, including my reconsideration request document—see Appendix B, and decided the evidence did not support a complaint of discrimination based on race. The EEOC made no mention of my hostile work environment or hiring disparity claims.

I filed a civil lawsuit concerning the same EEO complaints and the Judge in that case came to the same conclusion as the EEOC when he granted the justice department a summary judgement. The summary judgement said my claims were not worthy of being presented to a jury. I requested to take my case to a jury because I knew a jury of my peers would have agreed my Army leaders were guilty of fostering a hostile environment based on race.

I filed my third complaint with the MSPB in 2014 in response to my leaders retaliating against me for disclosing that team leaders had circumvented my authority, improperly influenced source selection decisions, and removed several of my MSTC leadership and engineering duties. The MSPB viewed the evidence and my response to the Judge's Close of Record Order—see Appendix C. The MSPB Judge ruled I had not presented evidence substantiating my claims. He added that I had only offered bare allegations. Because I had submitted numerous email messages supporting my claims—not just bare allegations, I decided to file an appeal.

I unsuccessfully appealed to the Federal Circuit. To be successful on appeal I had to clearly show that the MSPB Judge abused his discretion or violated a law when he ruled against me. The Circuit Judge was not persuaded by my arguments. Therefore, he ruled against me. Because I felt strongly that my leaders had retaliated against me and the proof was in the submitted evidence, I attempted to appeal my case to the U.S. Supreme Court.

U.S. Supreme Court appeals require the submission of 40 copies of your documents. If the Clerk believes an error exists in any document you submitted, your 40 copies will be rejected and returned to you—luckily, with your submittal fee—roughly $500. My documents were returned to me and I was given a few days to resubmit. I quickly learned why few cases are appealed to the Supreme Court. I decided against making the requested corrections, paying to reprint the 40 document packages, and resending the box of documents. I was certain the Clerk would find another issue with my submittal. I accepted that the case was over.

I filed my fourth complaint with the MSPB in 2017 in response to my leaders' continued retaliatory actions against me. Nettie removed me, without cause, from the Lead Engineer position the MSPB ordered I be returned to and she issued me unjustified low performance review scores. Similarly to my past complaints, the MSPB Judge ruled in favor of my leaders after reviewing my Closing Statements document—see Appendix D. The Judge ruled I had not sufficiently proven my leadership duties were taken away from me. He also ruled that the performance review element scores I complained about (scores for communication and teamwork) would not have led to a higher overall score if he agreed the scores should have been what I indicated I felt they should have been. Since the overall score would not have increased, he decided I did not have a valid complaint concerning my overall score. I unsuccessfully appealed the Judge's ruling to the Federal Circuit.

I filed my fifth complaint with the EEOC in 2017 after my resume was removed from competition concerning a promotion position. A last-minute resume requirement was added after the leader of the selection panel received my resume along with the other resumes that were approved for the competition by a neutral organization. I submitted my Closing Statements document to the EEOC for review—see Appendix E.

I filed my sixth complaint with the MSPB in 2018 in response to my leaders issuing me two reprimanding letters within two months and following up with the lowest performance review score I ever received in my professional career. The letters included false allegations concerning me sending union-interest email messages to my coworkers and verbally threatening my Supervisor. After I submitted my Closing Statements document, the MSPB Judge ruled in favor of my leaders. I unsuccessfully appealed the MSPB Judge's ruling to the Federal Circuit via an Informal Brief—see Appendix F.

I filed my seventh complaint with the EEOC in 2018 in response to my organization's Attorney releasing my unredacted EEOC complaint from 2011 (my first EEOC complaint) to several of my coworkers. One of those coworkers—Tammy, who worked with Randy to have me removed from the Bradley team—uploaded my complaint in a folder she labeled "Harroll Ingram Litigation" to the organization's common network drive where all employees (roughly 1,000 people) could view it. The complaint included my social security number, home address, telephone numbers, a characterization of my mental state, and charges I leveled against several organizational leaders. I submitted a Positional Statement to the EEOC for review—see Appendix G.

Despite my numerous pleas to federal authorities and local news stations for help, the authorities and news station personnel failed to properly investigate my claims and take actions to address the wrongdoings and wrongdoers. I used every reasonable legal option available to me to gain assistance in stopping the illegal actions of my coworkers and leaders. However, I was not successful. My coworkers and leaders were able to continue their illegal activities. Ultimately, I was forced to accept that I had no one to tell. I was not going to find help via the processes designed to protect federal employees from the retaliatory actions of their leaders. My plight gave me a new perspective of and level of empathy for people in my position who used alternate means to respond to their harmful unethical leaders.

Stop Trying To Make Cats Bark

I learned valuable lessons as a result of the numerous complaints I filed. In addition to accepting that justice isn't blind, I learned that we cannot force people to be what we believe they should be. I cannot force an unethical person to behave ethically. In my youthful days, my friends and I used an analogy to explain the concept. We used to say "you can't turn a hoe' into a house-wife." At my job, I wrote a message to myself at the top of my white-board. It read, "STOP TRYING TO MAKE CATS BARK!!" I used the message to remind me to disengage from disagreements by saying, "I understand your position" or "Maybe you are right". The expression also helped me accept that my position, which might appear to be correct today, might prove to be incorrect tomorrow. A lesson in humility and an important attribute of great leaders.

Leaders

Leaders create and share visions as well as motivate and inspire their followers to strive to reach those visions. Residing at the top of an organizational chart does not make one a leader. Although the chart position does inform of who in the organization holds the kite-strings (i.e., who is in control), it does not mean that person knows how to use those strings to make the kite dance. Just as skill is required of the kite flyer, a leader must be skilled to garner the necessary responses from her followers in an effort to meet the mission. For example, a leader who uses the skills of honesty and trustworthiness find less difficulty in convincing his followers to adopt his vision and strive to reach any leader-specified goals. A leader who empathizes with her followers facilitates the sharing of information that would not normally be shared between teammates. Thereby, creating a family-like bond and work environment. Great leaders protect their employees and will not hesitate to change strategic direction to avert any harm that might come to the team. Most importantly, good leaders do not blindly follow paths laid by predecessors.

Researchers of leadership theories have come to accept that leaders set the stage for how their followers will act. If a leader demands order, his followers generally perform in an orderly fashion. If a leader does not condemn a behavior, that behavior is often viewed as acceptable by her followers. Most of the leaders I've worked for during my DOD career have been unskilled in leadership. That resulted in followers who were deprived of opportunities to correct bad behaviors that would likely result in them also becoming unskilled leaders.

Most of the military and civilian personnel who were designated as leaders on my teams behaved as managers. Rather than blaze new paths for organizations, managers follow paths created for them. Managers generally focus on tasks—not missions or visions. Many of the managers placed in leadership roles in my organization were home-grown. They were born and/or reared in the Central Florida area. As research has shown, many of the children who were schooled in Florida were advanced to higher grade levels before they retained the knowledge offered at the lower grade levels. If those children, once they became adults, were hired and placed in leadership roles at my organization, there's no wonder why the organization has suffered from numerous poor leadership decisions in the past. The nepotism at my organization facilitated generations of poor leaders who carried on the aims of their ancestors. Many of my leaders, it seemed, chose to continue the fratricidal activities of the leader they replaced rather than challenge the status quo.

Fratricide is defined as destroying members of your own team. Leaders and other workers in federal and civilian organizations often take actions against their coworkers that result in the destroying of their coworker's reputation and career. Seeking assistance from outside organizations in rectifying the fratricidal actions taken by leaders against followers is usually unsuccessful and sometimes results in a re-victimization of the complaining (i.e., help seeking) employee. One organization—the MSPB, as I mentioned earlier, rules against the help-seeking employee over 90% of the time. What I viewed as the incompetence of the MSPB Judges I encountered left me wondering if seeking assistance from a higher authority might be beneficial.

I sent a letter to the leader of our country informing him of the MSPB's less than 10% rate of assisting civilians who sought help concerning the improper actions of their agency leaders. I informed that the MSPB's admission that they rule against the person disclosing wrongdoing in more than 90% of their cases signified a serious problem. It stands to reason that 90% of the employees cannot be wrong. My full letter to President Obama is included in Appendix H. Although I expected meaningful action, or at least a thoughtful response as a result of my letter, I only received a form-letter response thanking me for my concern. The response left me questioning the leadership skills of our Commander-in-Chief. It became crystal clear to me that help did not exist—the cavalry wasn't coming.

Justice For Everyone

Justice is defined as fairness, moral rightness, and a system of law in which every person receives his due. The founding fathers of the United States discussed justice in the country's constitution. The 14th amendment informs that no law shall be enforced that deprives any person of equal protection. The U.S. Supreme Court's motto is "Equal Justice Under Law" and their justices take an oath swearing to administer justice without respect to persons and to honor the rights of the poor and rich equally.

My federal leaders and I pledged allegiance to the same flag wherein we both finished by saying "and justice for all" Evidently, my leaders were not sincere. Justice at my organization was never applied equally. People like me who were not in the clique or who questioned orders that appeared to violate laws, rules, or regulations were not afforded their equal rights. The federal judiciary tends to back federal leaders who violate employee rights. Some laws are written to ensure federal leaders are protected from employee lawsuits.

The Westfall Act, also known as the Federal Employees Liability Reform and Tort Compensation Act, precludes federal employees from being sued for claims arising under state tort law if they were acting within the scope of their employment. For example, if I sued Judas, my second level Supervisor, for defamation when he issued me the reprimand letter full of lies and read it aloud with another person in the room, my organization/agency likely would have used the Westfall Act to protect him. They could have responded by saying Judas was acting within the scope of his employment when he issued me the reprimand letter and the Judge would likely have dismissed my case. Judas would have been emboldened to continue issuing unjustified reprimands filled with lies to employees who did not blindly follow orders. Judas, and leaders like him, use their positions of power as well as legal escape hatches like the Westfall Act as weapons against their employees. Just as many police officers cower once they are separated from their weapon, many leaders show their true selves (i.e., their weaknesses) when they are removed from their positions of power. Those organizational leaders who ruled their employees with an iron fist cower and beg through their tears for leniency once prosecuted for the crimes they committed. Those cats show how soft (and pink) they really are.

Federal employees and other citizens of our nation deserve to be protected equally by the law. Judges who use tools like the Westfall Act to issue decisions that re-victimize complainants contradict their oath to administer justice equally. Organizational leaders who react to complaints by protecting the accused do great harm to their organizations. Recent judicial decisions support the notion that justice is applied unevenly by our federal judiciary.

The FBI completed an investigation concerning a wealthy financier named Epstein who reportedly had a relationship with a high-ranking official in the Israeli Government—a U.S. Government ally. Epstein had also been accused of multiple sex crimes with minors. Although the results of the investigation led to an indictment, a U.S. Attorney granted Epstein immunity from prosecution. The Attorney later said he was told Epstein belonged to intelligence and that he (i.e., the Attorney) should leave it alone because it was above his pay grade.

Leaders at my organization, advised by their legal staff, investigated an allegation that a Colonel sexually harassed one of his employees by taking under-the-table photos of her during meetings. Despite the results of the investigation showing the Colonel was guilty, my leaders shielded the Colonel from punishment and assisted him in quickly retiring to preserve his financial status. Many federal leaders are unethical cats who cannot and will not ever bark.

Conclusion

The world and its people are often nothing like we expected them to be. I expected to enter the workforce and earn my way through the ranks fair-and-square. Although I was aware of racist-based activities aimed at blacks by privileged people, the privileged leaders I had (prior to entering the federal workforce) at least showed appreciation for a job well done. I did not find that to be the case in the federal government. My federal leaders went to great strides to avoid recognizing my accomplishments and to block my professional progression. Many of them ignored the law if it meant they would have to take my side in a dispute against a white person. In their view, apparently, I had no right to dispute anything said or done by a white person. In the distant past, as we all know, when a black man (i.e., a slave) was docile and simply did what he was told, he was viewed favorably by his master. However, once that black man began to develop his own mind and questioned the actions of his master that black man was immediately viewed as a threat and was attacked. In my view, relative to the mentality of most white people concerning race relations, not much has changed over the past centuries.

The federal government, over a century ago, abandoned southern African Americans who were being terrorized by racists white people for fighting for racial order. The government simply ignored its own laws when it came to supporting lawful blacks against unlawful whites. President Woodrow Wilson, in 1913, formally authorized the segregation of the federal government. Wilson's administration personnel reported that forcing white and black employees to work together and share facilities was intolerable. Thirty-five years later, in 1948, President Harry Truman passed two executive orders reversing Wilson's authorization and desegregating the Army and the federal government. However, the uncontrolled racist urges of many white federal workers led to black federal workers being relegated to a lower-working-class status. Those out-of-control white federal workers operated in the spirit of their 1880s ancestors who treated recently-freed blacks badly (i.e., repeatedly stopped the progressive actions of free black people). In many cases, those blacks petitioned the courts to be re-enslaved. Those blacks accepted that they were treated better while they occupied a lower caste—while they were enslaved.

Army Private Henry Johnson, a black man from Albany New York, served on the front line in World War I. Although Private Johnson bravely fought off over 20 German Soldiers, saved the life of his observation-post mate (Private Needham Roberts), saved the lives of French Soldiers who could have been killed by the approaching Germans, and was awarded France's Croix de Guerre medal for performing heroic deeds in combat, Private Johnson was not recognized for his war-time actions by the United States Army. In fact, Johnson reported that his white brothers-in-arms in the United States used racial slurs against him and did not want to serve in the trenches with him. The United States eventually loaned Private Johnson and others in the black regiment in which he served to the French Army for the duration of the war. When France recognized Johnson with a medal for his valiant service, the United States Army admonished France for the recognition. The United States told France Johnson and others like him might return home expecting to be treated as though they were equal to whites. Private Johnson's ordeal, although it took place over 100 years ago, parallels what many black men experience while working for the United States Army today. To right a century-long wrong, U.S. President Barack Obama presented a posthumous Medal of Honor to a relative of Private Johnson (on June 2, 2015) for "conspicuous gallantry and intrepidity at the risk of his life above and beyond the call of duty."

Becoming cognizant of historical events related to the inhumane, and to a lesser extent, unprofessional treatment my people have been forced to endure at the hands of white people has often left me questioning whether or not any good white people exist. Of course, the usual answer is yes. And I still agree with that answer. However, I must quantify my agreement by saying not many. I learned of one sure way to tell if a white person is good. Judge them by their decisions. If you know of a situation where a black man disagrees with a white man and that black man's position is clearly correct, present that dilemma to a white person and watch him decide. If the dilemma-settling white person responds by agreeing the black man's position is correct, maybe you have found a good white person. If that white person tries to find ways to credit the clearly wrong white man in the dilemma or discredit the clearly right black man, chances are your quest to find a good white person continues.

More than one hundred years after President Harry Truman integrated the federal government, most blacks at my organization continue to suffer from the separatist practices that embody the spirit of the pre-Harry Truman era. We suffer from the illusion of inclusion as we are treated as though we are forever in a lower caste. I witnessed separatist practices, on one occasion, as I read the confidential email messages, concerning me, sent between my coworkers. I gained access to those email messages via legal discovery requests. One message referred to me as an ass while another message defamed my character in freestyle rap lyrics.

I could not help but wonder how many other black federal workers were being victimized by the actions of their leaders and other coworkers due to uncontrolled racist urges. I suspect too many to count. Of those, I wondered how many were re-victimized by the system supposedly put in place to assist federal victims. Maybe that Navy Pilot back in Maryland who told me I would not like Florida because of the heat was trying to warn me about the hellish nature of some Floridians. When he added that I would have to shower three times a day I suppose he was referring to the dirty tricks played down south. I suspect the problem is much bigger than just what I and others have witnessed at my organization. The problem is likely systemic throughout the federal government.

My episodic career taught me that outward appearances and gestures are usually not trustworthy. My old mentor and friend, Reganol, told me years ago no one loves me but my momma. He also stressed that, like him, I was behind enemy lines at work. Although initially it was hard to accept, I eventually realized he was right. I also accepted that I was delusional in believing racial integration to promote equality on a professional level was a successful endeavor. Also, I was enlightened to the fact that using the go-along-to-get-along approach only temporarily shielded black people from the reality that your professional presence is not desired in predominantly white organizations.

Remaining focused on what matters most—family—is paramount. I realized the importance of dedicating more of my heart and soul to my blood family than to my "work family". I also learned to focus more on my health. There was nothing at my job of a higher priority than my family and my health. I realized I had to allow most of the actions my organization's leaders and their lackeys took against me to go unopposed (i.e., I had to forego responding to every negative action). Those actions were really inconsequential. However, my silence was/is not my surrender. I reserve the right to fight when necessary. I vow to stand (never again sit) when harmful actions are taken against me or my family.

My focus on excelling in my career shifted to helping others recognize when it's time to stand. Especially against "leaders" who unlawfully attack without guard rails. Booker T. Washington once wrote, "Success is to be measured not so much by the position that one has reached in life as by the obstacles which he has overcome while trying to succeed." Although the law says federal employees (including Supervisors) who engage in misconduct shall be disciplined, I and many others at my organization suffered years of discrimination and retaliation at the hands of our leaders without ever seeing those leaders disciplined for their actions. Even still, I refuse to cow-tow to offenders who behave as though they are above-the-law while the law enforcers allow them to sustain that behavior. I refuse to accept placement in a lower caste. I choose to forever fight against those who bully me—even at work. A great man said we must decide whether we will use our wishbone or our backbone. I refuse to rely on wishes, hopes, and dreams that equality for human beings like me will be attained. Instead, I will stand and demand equality with the understanding that equality should be a universal goal. I know the universe/karma ensures equality/balance. As I was taught at NSU, every action has an equal and opposite reaction—karmic retribution is always eminent!

Anyone who is being professionally mistreated like I have been should remember one piece of advice. Do not expect your offender (i.e., that cat) to change her behavior. Remember; Cats Don't

Bark. And do not experience the mistreatment silently. HOLLER! As with any criminal-low-life, once your offender knows you will reveal what he is doing, he will seriously consider stopping his actions against you. Once he knows someone with power is listening to your complaints, he will leave you alone.

Writing this book served as my therapy as I was forced to address the racist-based professional bullying I experienced during the last decade of my federal career. I also wrote numerous articles that cryptically covered the atrocities that took place in my organization. Hopefully, my story will be helpful to others who find themselves in a similar position to the one in which I suffered. If so, that person needs to know family will get him through the maze and he will be able to look back and see the glory of the learning experiences that hopefully facilitated his spiritual growth. In time, the proverbial bitter lemons he was force-fed will become sweet lemonade. In the words of Dr. Martin Luther King Jr, "Only when it is dark enough can you see the stars."

Appendix A

PETITION FOR BOARD REVIEW (2010)

HARROLL INGRAM

████████████████

The Clerk of the Board
Merit Systems Protection Board
1615 M Street, NW Washington, DC
20419

HARROLL INGRAM   Docket No. AT-1221-09-0874-B-1
Appellant, Pro Se, v.

DEPARTMENT OF THE ARMY (PEOSTRI)   Date: 15 September 2010
Agency.

APPELLANT'S PETITION FOR BOARD REVIEW OF INITIAL DECISION

The Appellant submits this petition for a Board review of the initial decision to deny the Appellant's

request for corrective action because the Administrative Judge believed the Appellant failed to show,

by preponderant evidence, that he disclosed a violation of any law, rule or regulation, gross

mismanagement, or abuse of authority. In this Petition for Review, the Appellant includes evidence

supporting his claims that the initial decision was based on misunderstandings of facts. The

misunderstandings relate to the Administrative Judge's assertions that the Appellant advocated for the

event, the Appellant viewed the dispute as a

performance issue, LTC ████████ maintained the event could still take place with no photographs

or videotape, and the Appellant disclosed information that was already known. The Appellant is

hopeful the clarifications provided below will result in a complete review of the facts in the appeal

documentation and corrective actions that will make the Appellant whole.

Background

The Appellant served on the Medical Simulation Training Center (MSTC) team as a Lead Engineer and Assistant Deputy Product Manager (dual roles). The Appellant was told by LTC ████████████, the Product Manager for the MSTC team, to lead a group of people on a MSTC product demonstration trip in Colorado. The group of people the Appellant was asked to lead included Government personnel and representatives from a contractor company that offered medical training products touted as better than the products the MSTC currently used. When the group expressed a plan to photo the MSTC products and Soldiers using the products, the Appellant questioned the legality of the plan. The Appellant obtained legal advice from the Agency's legal office and reported to LTC ████████ that the group's plan would violate a law. When LTC ████████ decided to continue the event against the suggestions of the Agency's legal office, the Appellant reported the imminent illegal activities to Mr. ████████ (the Appellant's Supervisor) and his desire not to take part in illegal activity as ordered by LTC ████████. When Mr. ████████ indicated that LTC ████████ had a right to overrule legal and that the Appellant must take part in the event, the appellant sought assurances from ████ and ████ Supervisor (Mr. ████████) that the Appellant would not be held accountable for violating any laws while taking part in the event. After not receiving the assurance and receiving a command-wide message from the Assistant Secretary of the Army informing that violating the subject law would lead to prosecution, the Appellant refused to take part in the event. Subsequently, the Appellant was ostracized from MSTC activities, eventually removed from the MSTC program, and given a lower than earned performance evaluation score—thereby harming the Appellant's opportunity for a promotion.

The Appellant filed a complaint alleging retaliation for whistleblowing activities with the Office of Special Counsel (OSC) alleging that as a result of his disclosures of LTC ████████

(attempted) violation of regulations and abuse of authority, he suffered negative personnel actions. The OSC ultimately rejected the Appellant's claim and issued him a closure letter terminating its inquiry into his complaint.

The Appellant then appealed to the Board and the Administrative Judge dismissed the appeal for lack of jurisdiction. The Administrative Judge concluded that the Appellant did not make a protected disclosure because (1) the Appellant's disclosures were part of his normal duties, (2) he did not make the disclosures to anyone with authority to correct them, (3) the disclosures were made to the wrongdoers, and (4) the purported disclosures were simply disagreements with management.

The Appellant filed a petition for review of the Administrative Judge's initial decision and the resulting ruling was that the Board had jurisdiction over the Appellant's appeal. The appeal was remanded for further adjudication.

The Appellant filed a response to the Board's Close of Record Order on July 27, 2010 offering statements and evidence concerning the Appellant's whistleblowing claims, the Appellant's protected disclosure, and the adverse personnel actions taken against the Appellant as a result of the protected disclosure. The Appellant also included information concerning damages in his response. The Administrative Judge denied the Appellant's request for corrective action because the Administrative Judge believed the Appellant failed to show, by preponderant evidence, that he disclosed a violation of any law, rule or regulation, gross mismanagement, or abuse of authority. The Appellant, as a result, filed this petition for review of the Administrative Judge's initial decision.

Initial Decision Analysis

The Administrative Judge based his corrective action denial on his view that the Appellant advocated for the event. The Administrative Judge accepts that the Appellant wrote to his manager that the event would be allowable if photos and videos were prohibited. Since the Appellant's advocacy stipulated that photos and videos be prohibited but LTC ████████ directed that the event continue with photos and videos (See tab 1, page 1), the Administrative Judge erred in basing his corrective action denial on the view that the Appellant and LTC ████████ advocated the event be carried out in accordance with legal opinion. The Appellant never advocated for photos and videos during the event. The Appellant believed, as the record shows, that carrying out the event with photos and videos violated a law, rule, or regulation and therefore his disclosure was protected in accordance with 5 U.S.C. § 2302(b)(8).

The Administrative Judge based his corrective action denial on his view that the Appellant viewed the dispute over whether or not the event should be carried out against the organization's legal opinion as a performance issue. The Appellant never indicated that he viewed the dispute as a personnel issue. The Appellant repeatedly showed that he believed LTC ████████ order would lead to a violation of law, rule, or regulation and the Appellant refused, out of fear of prosecution, to be forced to violate the law, rule, or regulation.

The Administrative Judge based his corrective action denial on his view that LTC ████████ maintained the event could still take place with no photographs or videotape, in an effort to accommodate the legal opinion. LTC ██████ directed that the event continue with photos and videos (See tab 1, page 1). Therefore, LTC ██████ maintained the event could still take place in spite of the legal guidance. The Appellant's Supervisor indicated that he recognized that LTC ████████ was overruling the legal department's guidance (See tab 1, page 2). LTC ████████ blatant disregard for the law as explained by the organization's legal department supports the

Appellant's claim that his disclosure to his Supervisor concerning the eminent law violation was protected in accordance with the Whistleblower Protection Act. The Administrative Judge erred in basing his corrective action denial on the view that LTC ███████ maintained the event could still take place with no photographs or videotape, in an effort to accommodate the legal opinion.

The Administrative Judge based his corrective action denial on his view that the Appellant disclosed information that was already known. The Appellant's Supervisor did not have knowledge of the event or the eminent violation of law, rule, or regulation prior to the Appellant's disclosure. Nothing in the record supports the Administrative Judge's view that the Appellant disclosed information that was already known. The Administrative Judge cited Meuwissen v. Department of the Interior to support his view that corrective action should be denied because the Appellant disclosed information that was already known. In Meuwissen v. Department of the Interior, the Judge found that Meuwissen failed to identify any law, rule, or regulation that was violated. In this case, the Appellant identified to his Supervisor the law, rule, or regulation that would be violated if the event was carried out. The Administrative Judge erred in basing his corrective action denial on the view that the Appellant disclosed information that was already known and cited an irrelevant case decision.

The Administrative Judge included, in his initial decision document, an assumption that the information in the memo from the Assistant Secretary for the Army was already known by the Appellant. The memo included information that was not a part of the response the Appellant received from the organization's legal department. The memo from the Assistant Secretary for the Army mentioned that Army civilians could be prosecuted for giving the appearance of giving preferential treatment to a non-Federal entity. Although a copy of the memo was not given to the

OSC, a copy was given to the Appellant's Supervisor to support the Appellant's belief that LTC ████████ order would lead to a violation of law, rule, or regulation. The Appellant did include a large quote from the memo in his filings to the OSC (See tab 1, page 3). The Administrative Judge erred in basing his corrective action denial on the view that the Appellant could not have reasonably believed that he disclosed a violation of any law, rule, or regulation to his Supervisor. The Administrative Judge also erred in basing his corrective action denial on the view that a disinterested observer, with knowledge of the essential facts known to and readily ascertained by the Appellant, could not reasonably have concluded that the disclosure evidenced wrongdoing of the type contemplated by the Whistleblower Protection Act.

Conclusion

The Administrative Judge denied the Appellant's request for corrective action because the Administrative Judge believed the Appellant failed to show, by preponderant evidence, that he disclosed a violation of any law, rule or regulation, gross mismanagement, or abuse of authority. The Appellant believes the Administrative Judge's decision to deny the Appellant's request for corrective action was based on misunderstandings of facts. The Appellant asks that the Board accept his request for review of the initial decision, consider the factual corrections (substantiated by the record) offered by the Appellant, recognize that the Agency failed to demonstrate by clear and convincing evidence that it would have taken the same actions against the Appellant without the disclosure--as ordered by the board on May 7, 2010, and order the corrective actions requested by the Appellant in his response to the Board's Close of Record Order dated July 27, 2010.

Respectfully submitted,

Appendix B

RECONSIDERATION REQUEST (2015)

Harroll Ingram

██████████████

██████████████

Director

U.S. Equal Employment Opportunity Commission

Office of Federal Operations

P.O. Box 77960

Washington D.C. 20013

DA Docket No: ARCESAV11JAN04975 (Appeal No: 0120131291) Date: 4 June 2015

Dr. Harroll Ingram

Complainant, Pro Se.

v.

John M. McHugh, Secretary of the Army

Agency.

COMPLAINANT'S REQUEST FOR RECONSIDERATION IN RESPONSE TO THE EQUAL EMPLOYMENT OPPORTUNITY COMMISSION'S FINAL DECISION

The complainant submits this appeal reconsideration request concerning the EEOC's decision affirming the Agency's final decision related to the Complainant's allegation of unlawful employment discrimination on the basis of race (black—African American) in violation of Title VII of the Civil Rights Act of 1964. The Complainant includes arguments and evidence in this request establishing that the decision involved clear erroneous interpretations of facts and laws.

The Complainant is hopeful the EEOC will reverse its initial appeal decision and deny the Agency's final decision.

Background

The Complainant served as the Lead Engineer and Test Director on the Bradley Advanced Training System (BATS) project and the Bradley Conduct of Fire Trainer-Enhanced (COFT-E) project at the U.S. Army's Program Executive Office for Simulation, Training, and Instrumentation (PEOSTRI) command located in Orlando, Florida. Ms. ███████ served as the Lead Engineer and Test Director on the Bradley Conduct of Fire Trainer-Situational Awareness (COFT-SA) project, Mr. ███████ served as the Bradley team's Subject Matter Expert (SME), Mr. ███████ served as the Project Director (PD) for all Bradley projects, Mr. ███████ served as the Complainant's first level supervisor, Ms. ███████ served as the Complainant's second level supervisor, and Mr. ███████ served as the Complainant's third level supervisor. Mr. ███ was one of two contracted employees on the Bradley team. Mr. ███ had recently been given several logistics tasks because the team's contracted Logistician was fired by the company for which he and Mr. ███ worked. Ms. ███ and the Complainant were both assigned (matrixed) to the Bradley team by the engineering department and both reported directly to their first level supervisor—Mr. ███████

The Complainant reported several incidents, including harassing and disrespectful email messages and other activities perpetrated by Mr. ███ and directed at the Complainant, to his supervisors. The Complainant's supervisors failed to take prompt and effective remedial actions to resolve the hostile work environment. The Complainant's leaders chose to remove the Complainant from his Lead Engineer and Test Director positions and place him in an unclear position on a non-existent Bradley Fire Support Team (BFIST) project. After the Complainant revealed to his leaders that he was aware that the BFIST project did not really exist, the Complainant's leaders moved him to another program where he served in a lower capacity (non-Lead Engineer). The Complainant subsequently filed an EEO complaint, completed the pre-complaint process, participated in the Fact Finding Conference (FFC), received an ROI from the Investigator, received a final decision from the Agency, and appealed the Agency's decision to the EEOC. The EEOC issued a decision on the appeal affirming the Agency's final decision. This reconsideration request reveals errors in the EEOC's interpretation of material facts and laws. The Complainant believes reconsideration will support a reversal of the EEOC's decision to affirm the Agency's final decision.

EEOC Decision Analysis

The EEOC's decision to affirm the Agency's final decision was based on erroneous interpretations of material facts. The EEOC relied on the Agency's misinformation that Mr. ███████████ (support contractor) gunnery experience and logistics experience made him indispensable to the COFTE team and led to the removal of the Complainant from the team. The case record shows the Complainant's leaders (Mr. ███████████—PD and Mr. ███████████— Chief Engineer), along with the Complainant's co-worker (Ms. ███████████) testified under oath they decided to remove the Complainant from the COFT-E project and put him on the BFIST project because the BFIST project was something the Complainant could execute on his own without IPT involvement. See pages 464 – 466 of the ROI. The Complainant's leaders also testified they removed the Complainant because Mr. ████ was more valuable to the team. See pages 415, 416, and 419 of the ROI. The EEOC's decision shows the Agency stated the hostile environment between the Complainant and Mr. ████ was the reason they removed the Complainant. The various reasons given by the Complainant's leaders are contradictory and support the Complainant's allegation that the reasons given were pretextual and given to hide the racially discriminatory reason for removing the Complainant. The Agency did not give a reliable reason for removing the Complainant from his leadership position and placing him in a lower level position negatively affecting his ability to meet the Agency's performance goals and his ability to populate his resume with leadership experience in an effort to gain a promotion.

The EEOC failed to consider material facts when it ignored the Agency's failure to consider the Complainant's expertise when alleging the support contractor, Mr. ███, was more valuable to the team. According to sworn testimony, the BATS project led by the Complainant within the hostile environment perpetrated by the support contractor and sustained by the Complainant's leaders was remarkably successful. See pages 462 of the ROI. The Complainant has extensive expertise in several areas (over 17 years of acquisition experience in modeling and simulation, over 20 years of testing experience, multiple certifications in testing and systems engineering, served on multiple PEOSTRI programs as Lead Engineer and Test Director, holds an Associates degree in computer networking, a Bachelors degree in engineering, a Masters degree in management, and a Doctorate in organizational leadership). The support contractor's experience concerned an older version of the Bradley vehicle—not the version that was the subject of the project. The support contractor was not really an expert on the BATS or COFTE simulator operations that were being acquired by the team. The true BATS and COFTE experts (2 of them) were sent to the team by the U.S. Army—FT Benning. The support contractor also was not a logistics expert. He had recently adopted the logistics function after the team's Logistician was fired from his company. The government team's contract with the company that provided Mr. ███ required the company to replace the fired Logistician. The company asked Mr. ███ to take over the function. Mr. ███ was referred to as an acting Logistician by the PD on page 451 of the ROI. The Agency's decision that Mr. ▮▮▮ (the support contractor whose employer had a business relationship with the government) was more valuable to the team than the Complainant (the government employee who was thoroughly trained to perform project duties in the interest of the taxpayer), was unreliable. The PD informed, on page 451 and 452 of the ROI, government employees and contracted employees were treated as equals on the team. Equating Mr. ▮▮▮ with the Complainant (placing the support contractor ahead of the Complainant government employee in this case) put the support contractor in a situation where it could be assumed he was an agency employee or representative and arguably violated regulations concerning giving contractors inherently governmental functions. See Federal Acquisition Regulation § 7.503. A fair and reliable comparison of the support contractor's value to that of the Complainant must be carried out by qualified individuals—not Mr. ▮▮▮, Mr. ▮▮▮, or Ms. ▮▮▮. See Collier v. Budd Co., 66 F.3d 886 (7th Cir. 1995).

The EEOC misinterpreted material facts when it mischaracterized the essence of the Complainant's allegation by writing the Complainant alleged that his management chain created a hostile work environment by appointing a Lead Systems Engineer over him and by allowing a personality conflict between him and a support contractor to worsen which resulted in his removal as the Lead Engineer. The Complainant did not state or imply his removal resulted from a personality conflict. The Complainant informed that his removal from the leadership position resulted from unlawful employment discrimination on the basis of race in violation of Title VII of the Civil Rights Act of 1964. The conflict between the Complainant and the support contractor should have been resolved by the Complainant's leaders but the Complainant's leaders chose to ignore the Complainant's pleas for help and chose to secretly assist the support contractor in continuing his hostile actions against the Complainant. The Complainant's leaders (Mr. ▮▮▮

████—PD and Mr. ██████████—Chief Engineer) and his co-worker (Ms. ██████████) chose to unlawfully remove the Complainant from his leadership position and create a pretext (e.g., the support contractor was more valuable to the government's team than the government leader) to protect themselves from possible administrative/legal repercussions.

The EEOC misinterpreted material facts when it mischaracterized the crux of the Complainant's allegation by writing the crux of the Complainant's harassment allegation is that the PM and PD allowed the support contractor to ignore his prerogatives as the lead on the COFTE project and that they ultimately removed him from COFTE rather than remove a support contractor. The crux (i.e., the most important point) of the Complainant's allegations is that the PD—Mr. ████ ████, Chief Engineer—Mr. ██████████, and the Complainant's co-worker—Ms. ██████████ unlawfully conspired to (and did) remove the Complainant from his leadership position. The Complainant's leaders also created a pretext for the removal in an effort to protect themselves from possible administrative/legal repercussions.

The EEOC failed to consider material facts when it relied on the Agency's claim that the Complainant had to be removed, while the support contractor was retained, to stop the negative impact on the team. If the EEOC had considered the additional facts, not just the Agency's self-serving claim that the Complainant had to be removed, the Commission would have denied the Agency's final decision. One fact is the Agency (i.e., the Complainant's leaders) could have employed a less discriminatory alternative rather than remove the Complainant. The Agency could have contacted the support contractor's company to facilitate an improvement in the support contractor's behavior. Instead, the Complainant's leaders gave the support contractor advice to help him avert repercussions of his hostile/inappropriate actions against the Complainant. See pages 511 - 512 of the ROI. The Agency subsequently hired the support contractor and made him a government employee soon after the Complainant was removed from the team. See page 448 of the ROI.

The Complainant's leaders admitted under oath they felt the support contractor's actions were improper. See pages 417, 445, 473, and 487 of the ROI. The Complainant's leaders also admitted they never contacted the support contractor's company about his behavior. See pages 417 – 421, 445, 446, and 513 – 514 of the ROI. Improper actions/behavior by personnel of companies who have a contractual relationship with the government violate the stipulations of the government-private sector company contract. The Complainant's leaders could have/should have sought an intervention by the support contractor's company. The courts have found that the employer has an affirmative duty to investigate complaints of harassment and to deal appropriately with offending personnel. Failure to investigate gives support to the discrimination because the absence of sanctions encourages abusive behavior. See Zabkowicz v. West Bend Co., 589 F. Supp. 780, 35 EPD ¶ 34,766 (E.D. Wis. 1984).

The EEOC misinterpreted material facts when finding the Complainant has not presented any sworn statements from other witnesses or documents that contradict the explanations provided by officials or which call their veracity into question. The Complainant presented several sworn statements and documents that contradict the explanations of the Agency—as part of his Appeal of the Agency's Final Decision. The Complainant's leaders testified under oath they decided to remove the Complainant from the COFT-E project and put him on the BFIST project because the BFIST project was something the Complainant could execute on his own without IPT involvement. See pages 464 – 466 of the ROI. The Complainant's leaders also testified they removed the Complainant because Mr. ████ was more valuable to the team. See pages 415, 416, and 419 of the ROI. Additionally, the Complainant's leaders testified the hostile environment between the Complainant and the support contractor was only part of the reason the Complainant was removed. See pages 416 and 471 of the ROI. Another leader stated the Complainant was removed because of available resources and workload issues. See page 415 of the ROI. The Agency gave contradictory testimonies concerning why they removed the Complainant from his leadership position—clearly calling their veracity into question. The contradictory reasons also support the Complainant's allegation that the reasons given and relied on by the EEOC were a pretext for the true reason—discriminatory motivation.

The EEOC misinterpreted the law when indicating the Complainant needed sworn statements that contradict the explanations provided by officials or which call their veracity into question. In Linton v. City of Marlin, 25 F.3d 706 (5th Cir. 2001), unsworn statements submitted to the EEOC were deemed admissible because they could be otherwise authenticated with circumstantial evidence. The un-rebutted email messages submitted by the Complainant should be admitted as evidence along with the sworn testimony of the Complainant's leaders to support a denial of the Agency's final decision.

The EEOC misinterpreted the law when indicating the reason for affirming the Agency's decision was because none of the documents or testimony provided showed any indications of a racial animus on the part of the PM or PD. The law does not always require a clear discernment of racial animus from submitted documents and/or testimony to support a finding of discrimination. The law provided, in Texas Department of Community Affairs v. Burdine, 450 U.S. 248 (1981), that discriminatory intent can be inferred from disparate treatment. Evidence of disparate treatment may include proof that other, similarly-situated, employees not in the protected class received systematically better treatment. Marshall v. American Hospital Assoc., 157 F.3d 520 (7th Cir. 1998). The Complainant complained that his leaders treated him differently than his equally qualified peers—other Lead Engineers. The sworn testimonies of the Complainant's leaders support the allegation. Mr. ▮▮▮▮ testified that he could not recall removing any other Engineer from a team because they had difficulty working with team members—see page 428 of the ROI. Mr. ▮▮▮▮ testified that he could not recall any Engineers, other than the Complainant, being removed from a team because they had difficulty working with team members—see pages 457 and 458 of the ROI. Ms. ▮▮▮▮ testified that she could not recall of any other Engineer who was removed from a program because he/she did not get along with someone else on the team—see page 477 of the ROI. Mr. ▮▮▮▮ testified that he could not recall any other Engineers being removed from a program because of complaining about disrespectful actions of contractors. Mr. ▮▮▮▮ also testified that he could not recall any other Engineers being removed from their position because of emails or interactions with contractors—see pages 504 and 505 of the ROI. Mr. ▮▮▮▮ (the Engineer the Complainant replaced) had team issues with Mr. ▮▮▮▮. However, Mr. ▮▮▮▮ (a white former team employee) was not removed from the team—see page 14 of the ROI. Mr. ▮▮▮▮ was an equally qualified peer (similarly situated) of the Complainant and he had the same Supervisor (Mr. ▮▮▮▮), but was treated more favorably. See Little v. Ill. Dep't of Rev., 369 F.3d 1007, 1012 (7th Cir. 2004). The Complainant provided proof that his leaders have a history of disparate treatment against African Americans by showing a hiring rate of African American Engineers below 2%—statistically more difficult than pure chance and substantiating a disparate treatment finding. The EEOC has found if an organization's hiring rate of a particular group made it statistically more difficult than pure chance for a member of a certain group to get a job, then the hiring rate could be reasonably viewed as evidence that the selection system was systematically screening out members of that social group.

The Complainant's leaders, as the record substantiates, removed the Complainant from several leadership positions greatly harming his ability to gain experiences to show qualifications for promotions. The MSPB found, in case docket number AT-1221-09-0874-B-1, that the Agency unlawfully removed the Complainant from a leadership position and ordered the Agency to return the Complainant to the position. The Agency has repeatedly subjected the Complainant to unfair treatment and retaliation as shown in MSPB case docket number AT-1221-14-0725-W-1 (Federal appeal number 2015-3110) and OSC case number MA-15-3740. The Investigator in this EEOC case jotted a word in the margin of the ROI—page 8. That word was "reprisal". Apparently, the EEO Investigator wondered whether or not retaliation played a role in the actions the Agency took against the Complainant.

The EEOC misinterpreted material facts when finding that the Complainant failed to satisfy his burden to prove the existence of a discriminatory motivation on the part of the PM, PD, or any other individual within his chain of command or on his team in connection with any of the matters at issue in his complaint. The Complainant proved the existence of a discriminatory motive on the part of his leaders using sworn testimony that gave several conflicting reasons for his removal from the leadership position. The Complainant also proved his leaders treated him differently than his equally qualified white peers. The conflicting testimony and disparate treatment supports the allegation that the Agency failed to articulate a legitimate non-discriminatory reason for the Complainant's removal. The Agency failed to meet its burden of showing by clear and convincing evidence that the Complainant would have been removed from his leadership position absent the discrimination. See Eyslee v. Dep't of the Treasury, EEOC Appeal No. 0720100050 (December 7, 2011). The Complainant, as a result, satisfied his burden of proving the existence of a discriminatory motive on the part of his leaders.

Conclusion

The EEOC relied on the Agency's misinformation when it issued its decision to affirm the Agency's final decision. A review of the investigator's report and the full record will support a realization that many of the Agency's statements were contradictory and not reliable.

The Agency misinformed that Mr. ███ gunnery experience and logistics experience made him indispensable to the COFTE team. The record supports that Mr. ███ was not really an expert on the BATS or COFTE simulator operations that were being acquired by the team. The true BATS and COFTE experts (2 of them) were sent to the team by the U.S. Army. Mr. ███ also was not a logistics expert. He had recently adopted the logistics function after the team's Logistician was fired from his company.

The Agency misinformed, under oath, that they decided to remove the Complainant from the COFT-E project and put him on a non-existent BFIST project because the BFIST project was something the Complainant could execute on his own without IPT involvement. Later, the Agency stated they removed the Complainant because Mr. ███ was more valuable to the team-- an allegation that was not substantiated in the record. The Agency also claimed resources and workload issues led to the Claimant's removal.

The Agency claimed the Complainant was poisonous to the team, but admitted the Complainant led the BATS project remarkably well--despite the hostile environment. The BATS was reported to be the best version ever delivered.

The PD—Mr. ███████, Chief Engineer—Mr. ███████, and the Complainant's co-worker—Ms. ██████ unlawfully conspired to (and did) remove the Complainant from his leadership position. The Complainant's leaders also created a pretext for the removal in an effort to protect themselves from possible administrative/legal repercussions. Although they admitted under oath they felt Mr. ███ actions were improper, the Complainant's leaders testified they never contacted the support contractor's company about his behavior.

The EEOC misinterpreted material facts when finding the Complainant had not presented any sworn statements from other witnesses or documents that contradicted the explanations provided by officials or which call their veracity into question. The Complainant presented several sworn statements and documents that contradicted the explanations of the Agency—as part of his Appeal of the Agency's Final Decision and repeated in this reconsideration request. The Agency's contradictory testimonies concerning why they removed the Complainant from his leadership position clearly called their veracity into question. The contradictory reasons also support the Complainant's allegation that the reason given and relied on by the EEOC were a pretext for the true reason—discriminatory motivation.

The EEOC misinterpreted the law when indicating the Complainant needed sworn statements that contradicted the explanations provided by officials or which call their veracity into question. Unsworn statements have been deemed admissible when they can be otherwise authenticated with circumstantial evidence. Several of the email messages in the record authenticate the Complainants statements and allegations of discrimination.

The EEOC also misinterpreted the law when indicating the reason for affirming the Agency's decision was because none of the documents or testimony provided showed any indications of a racial animus on the part of the PM or PD. The law does not always require a clear discernment of racial animus. Discriminatory intent can be inferred from disparate treatment. The Complainant substantiated disparate treatment by showing that other, similarly-situated, employees not in the protected class received systematically better treatment.

The EEOC misinterpreted material facts when finding that the Complainant failed to satisfy his burden to prove the existence of a discriminatory motivation on the part of the PM, PD, or any other individual within his chain of command or on his team in connection with any of the matters at issue in his complaint. The Complainant proved the existence of a discriminatory motive on the part of his leaders using sworn testimony that gave several conflicting reasons for his removal from the leadership position. The Complainant also proved his leaders treated him differently than his equally qualified white peers. The conflicting testimony, disparate treatment, and the failure of the Agency to articulate a legitimate reason for removing the Complainant from his leadership position legally support a finding of discriminatory motive. The Agency failed to meet its burden of showing by clear and convincing evidence that the Complainant would have been removed from his leadership position absent the discrimination. The Complainant, as a result, satisfied his burden of proving the existence of a discriminatory motive on the part of his leaders. The Complainant believes the EEOC, after reviewing the Complainant's request for reconsideration and the entire record, will find that the Complainant's request meets the criteria of 29 C.F.R. § 1614.405(c) and will reverse its earlier decision by denying the Agency's final decision.

The Complainant, as a result of the discriminatory actions and harassment, has suffered from depression, anxiety, stress, insomnia, difficulty concentrating, social isolation, damage to his professional reputation, withdrawal from relationships, loss of self-esteem, frustration, and humiliation. The Complainant also incurred administrative (postage) and legal expenses.

The Complainant argues that the EEOC's decision should be reconsidered because to allow it to stand will have a substantial impact on the practices of the Agency. If the Agency is allowed to arbitrarily remove government employees from leadership positions and place them in lower-level positions without cause, other than to claim a support contractor was more valuable, it will encourage agency leaders with similar circumstances to use the same strategy to keep offending contractor team members on teams.

The Complainant seeks to have the agency's engineering and program management departments directed to take annual conflict resolution and cultural diversity training, have the agency submit a formal written apology to the Complainant for subjecting him to a hostile and discriminatory work environment, and an award of $15,000 in non-pecuniary compensatory damages.

Harroll Ingram, Pro Se

<u>Appendix C</u>

<u>CLOSE OF RECORD ORDER RESPONSE</u>
<u>(2014)</u>

U.S. Merit Systems Protection Board

Atlanta Regional Office

Administrative Judge Richard W. Vitaris

401 W. Peachtree Street, NW, Suite 1050

Atlanta, GA 30308

IN RE: Docket No. AT-1221-14-0725-W-1   Date: 20 August 2014

Harroll Ingram

Appellant

vs.

Department of the Army (PEOSTRI)

Agency

APPELLANT'S RESPONSE TO THE U.S. MERIT SYSTEMS PROTECTION BOARD'S CLOSE OF RECORD ORDER

The Appellant offers the following statements and evidence concerning the Appellant's claim of retaliation for whistleblowing activity against the Agency. The Appellant also includes information concerning damages in this response.

Background

The Appellant served on the Medical Simulation Training Center (MSTC) team as a Lead Engineer and Assistant Deputy Product Manager (dual roles). The Appellant was responsible for all engineering duties to include writing specifications, statements of work, and other contractual documents as well as evaluating proposals and performing management duties relative to acquiring products, interfacing with customers, and representing the team at medical conferences around the country.

The Appellant, after engaging in whistleblowing activity and being punished by the Agency (i.e., being removed from his leadership position and given a lower then earned performance evaluation) as a result, filed a complaint with the Merit Systems Protection Board (MSPB). The MSPB found that the Appellant was unlawfully punished for whistleblowing activity and issued a final ruling in the subject case (AT-1221-09-0874-B-1) ordering the agency to increase the Appellant's performance review score and return the Appellant to his former leadership position.

The Appellant began to be subjected to retaliatory personnel actions soon after the MSPB's final order was issued and those unfair personnel actions have continued ever since. The Appellant's leaders, including those who either played a role in the initial whistleblower case or were aware of the initial case, retaliated against the Appellant and created and sustained a hostile working environment by scheduling the Appellant for overtime and refusing to approve overtime pay, blocking the Appellant's ability to perform his engineering duties, removing the Appellant from positions because he refused to join them in violating federal regulations and laws, taking actions to have the Appellant removed from his MSTC Lead Engineer position without cause, and issuing the Appellant a reprimand letter incorrectly accusing him of behaving in a manner that was detrimental to the organization's mission—the letter also included a threat of employment termination. The actions taken by the Appellant's leaders meet the knowledge/timing test verifying that the disclosure was a contributing factor in the decisions to take the personnel actions against the Appellant. The retaliatory actions were taken against the Appellant to punish him for using his right to complain and seek grievance and to show his co-workers what happens

to people who file complaints against the organization.

Significant Change in Duties

The Appellant's Project Director, Ms. ██████████, greatly reduced the Appellant's duties

and her work-related interactions with the Appellant (i.e., Ms. ██████ gave engineering duties

to and sought technical/engineering advice from non-engineers on the team—see Tab A, pages 1

through 43). As a result of Ms. ██████ actions, the Appellant suffered a significant reduction

in duties and a loss of opportunities to meet or exceed performance goals in an effort to gain a

promotion to a GS-14 level position. Ms. ██████ actions effectively relegated the Appellant

to a non-Lead Engineer status and left the Appellant in a position where he could no longer

perform the duties he formerly held as a Lead Engineer. Section 2302(a)(2)(A) of the

whistleblower protection act informs that a significant change in duties, responsibilities, or

working conditions constitutes a personnel action. The Appellant complained to his Supervisors

expecting them to take actions to stop Ms. ██████ abuse of power and unfair actions against

him. The Appellant's Supervisors failed to take any meaningful actions to assist the Appellant

(e.g., take actions to stop the power abuse and unfair actions).

Position Removal

The Appellant's Supervisor, Mr. ██████, and Project Director, Ms. ██████,

sought to have the Appellant removed, without cause, from the Lead Engineer position to which

he was returned by the MSPB. Ms. ██████ attempted to use pretextual reasons to influence the

Appellant's removal, while Mr. ██████ sought legal advice concerning removing the Appellant from one of the organization's lawyers (Ms. ████████). The removal attempts are substantiated in email messages—see Tab B, pages 1 through 11.

Reprimand Letter

The Appellant's Supervisors, Mr. ████████, and Mr. ████████, issued the Appellant a letter of reprimand for taking a restroom break before a teleconference officially ended. The Appellant's team leader, LTC ████████—a role-player in the initial whistleblower case, influenced the Appellant's Supervisors' decision to issue the reprimand letter. The letter was intended to modify the Appellant's behavior in the future and meets the corrective and covered personnel actions definitions of the whistleblower protection act. The Appellant met with his Supervisors to discuss the letter. The Appellant sought answers to several questions such as why didn't his Supervisor's seek his side of the story before issuing the letter, was the letter a reprimand, will they rescind the letter—after gaining a better understanding of what transpired—see Tab C, pages 1 through 9.

Refused Overtime Pay

The Appellant's team leader, LTC ████████, scheduled the Appellant to work 20 hours more than the normal 40 hours for one work week without overtime pay. LTC ██ also refused to approve compensatory time, an alternate way of rewarding overtime work, for the additional 20 hours—see Tab D, pages 1 through 7. Denying an employee an opportunity to earn overtime pay that would otherwise have been provided is a decision concerning pay and therefore is a personnel action. Although the Appellant refused to work the 20 additional hours without adequate pay, whistleblower reprisal can be found because it is retaliation just to threaten to take or not take a personnel action.

Promotion Opportunity

The Appellant's leaders created a Chief Engineer position on the Appellant's team and selected an Engineer from another team, without competition, for the new Chief Engineer position—see Tab E, pages 1 and 2. The selected employee, Ms. ████████, had no medical simulation experience. The Appellant was told, in 2008—prior to filing the initial whistleblower complaint, he was the logical person for the Chief Engineer position when the position opens. Therefore, the Appellant believes he would have earned the Chief Engineer position, based on his experience on the team, education, and leadership abilities, if a fair competition had been held. Further, The Appellant believes a fair competition would have been held if he had not filed the initial whistleblower complaint. Title 5 of the United States Code (5 U.S.C. § 2302(b)) informs that any employee who has authority to take, direct others to take, recommend, or approve any personnel action, shall not, with respect to such authority, grant any preference or advantage not authorized by law, rule, or regulation to any employee or applicant for employment (including defining the scope or manner of competition or the requirements for any position) for the purpose of improving or injuring the prospects of any particular person for employment. The Appellant believes his leaders' decision to create the Medical Simulation Chief Engineer position and select Ms. ████████, without competition, provided Ms. ████ an improper advantage and injured the Appellant's prospects for obtaining the Chief Engineer position.

Corrective Actions

The appellant requests that the MSPB order the following corrective actions:

- ☐ Order the agency to provide cultural diversity training for the Appellant's engineering leaders and the Appellant's medical simulation team co-workers.

☐ Order the agency to take steps to ensure the MSTC PD understands and follows DOD acquisition guidelines concerning use of personnel resources and adhering to DOD laws, rules, and regulations.

☐ Order the agency to rescind the LOC I was issued (dated April 16, 2012).

☐ Any other action the MSPB believes is necessary.

Conclusion

The Appellant engaged in whistleblowing activity and was punished by the Agency by being removed from his Lead Engineer position on the medical simulation team. The MSPB found that the Appellant was unlawfully punished for whistleblowing activity and ordered the agency to return the Appellant to his former Lead Engineer position. Within two months of being returned to the team, the Appellant began to be subjected to retaliatory personnel actions for filing the whistleblower case and to show the Appellant and others the negative repercussions of filing complaints against the Agency.

The evidence provided by the Appellant supports the claim that the Appellant's Project Director subjected the Appellant to power abuse and a significant change in duties. The evidence also supports the claim that the Appellant's Supervisors ignored the Appellant's repeated pleas for assistance in stopping the power abuse and worked in concert with the team's Project Director to find a way to remove the Appellant from the team to which the MSPB ordered he be returned. Because of the short span of time between the Appellant's return to the team and the start of the prohibited personnel practices taken by the Agency (people in leadership positions who knew about the initial whistleblower case) against the Appellant, a causal link can be reasonably made—the evidence gives rise to an inference of retaliatory motive. The Appellant requests the aforementioned corrective actions in hopes of being returned to the state he would have held if the Agency had not taken the retaliatory actions against him.

Respectfully submitted,

Appendix D

CLOSING STATEMENT (2018-A)

Judge Alexander G. Thompson

MSPB Atlanta Regional Office

401 W. Peachtree Street, NW, Suite 1050

Atlanta, GA 30308

HARROLL INGRAM   Docket No. AT-1221-17-0498-W-1

Appellant, Pro Se,

v.

DEPARTMENT OF THE ARMY   Date: 18 January 2018

Agency.

CLOSING STATEMENTS

I (the appellant), in the subject IRA Appeal hearing held on December 14, 2017, set out to show
that my Supervisors repeatedly retaliated against me by subjecting me to prohibited personnel
practices for making protected disclosures. I believe the evidence highlighted during the hearing
as well as what exists in the record substantiates the retaliation claims. This submittal documents
my closing statements concerning the hearing.

Protected Disclosures

I previously filed two separate appeals with the MSPB. The initial appeal (AT-1221-09-0874-B-
1) was filed in 2008 and concerned unfair and unlawful personnel actions taken against me (i.e.,
removing me from my Lead Engineer position and giving me a lower than earned performance
evaluation score) by my team leaders at PEO STRI in retaliation for informing agency leaders
that my team leaders, as well as my engineering leaders, were planning to violate multiple laws.
The MSPB found that my disclosures were protected per 5 U.S.C. § 2302(b)(8). My engineering
leaders who played key roles in the personnel actions that were taken against me were Mr.

█████████ and Mr. ████████. Mr. ████ and Mr. ████ served as supervisors of the two key engineering leaders who played key roles in the personnel actions taken against me in the current appeal.

The second appeal (AT-1221-14-0725-W-1) was filed in 2014 and concerned unfair and unlawful personnel actions taken against me (i.e., substantially taking away my Lead Engineer duties and issuing me a reprimanding letter incorrectly accusing me of behaving in a manner that was detrimental to the organization's mission as well as threatening me with termination) by my leaders at PEO STRI in retaliation for my whistleblowing activities related to the aforementioned 2008 case and for informing agency leaders that my team leaders, as well as my engineering leaders, were violating laws and acquisition guidelines such as approving government payments to vendors before the government received the vendors' products, giving my government engineering duties (inherent governmental functions) to support contractors, and scheduling me for 20-hours of overtime without formally approving me to work the 20 additional hours. The MSPB found that my disclosures were protected per 5 U.S.C. § 2302(b)(8). My engineering leader (immediate Supervisor) who played a key role in the personnel actions that were taken against me was Mr. █████████ (one of the accused in the current appeal). Mr. █████ Supervisor at the time of the second appeal was Mr. ████████—a key perpetrator in my initial appeal.

Prohibited Personnel Practices

The unfair and below expected performance review communication and teamwork scores given to me by Mr. █████ and Ms. ██████ directly affected (negatively) my earnings (i.e., resulted in a lower bonus for FY14 and FY15) thereby meeting the MSPB's definition of a prohibited personnel practice under 5 C.F.R. § 2302(a) which includes as a prohibited personnel practice "a decision concerning pay, benefits, or awards…." The agency's attorney questioned, during the pre-hearing conference, whether or not I was willing to waive any increase that results from increasing my performance review scores if a settlement was reached and Mr. ███ stated under oath, during the hearing, if he had issued me 84s for communication and teamwork (scores I indicated I deserved in my internal grievance) I would have received a higher bonus for FY14.

The removal of my Lead Engineer position and duties on the MSTC team relegated me to a non-lead position, put me back in an engineering position I worked hard over the past years to earn elevation from, and negatively affected my standing relative to future training/promotion opportunities thereby meeting the MSPB's definition of a prohibited personnel practice under 5 C.F.R. § 2302 which includes as a prohibited personnel practice "any other significant change in duties, responsibilities, or working conditions." The record shows I suffered a significant change in working conditions which constitutes a personnel action within the meaning of the WPA.

███████

Mr. ████████████ served as my Supervisor from FY12 through FY14. I reported several actions taken by my team leaders (e.g., Ms. ██████████) that I believed violated acquisition rules and laws governing how government employees should conduct government business. I reported the activities to several leaders who could and should have intervened to stop the activities (i.e., reported the activities to my Supervisors, Procurement Contracting Officer, and Military Officers). I eventually reported the disclosures to the OSC and the MSPB in support of case AT-1221-14-0725-W-1. Rather than stop the activities I reported, my leaders took personnel actions against me.

Mr. ██████ issued me low performance evaluation scores for two FY14 CCAS elements—communication and teamwork. After I formally (via an internal grievance) accused Mr. ██████ of issuing me low scores, he created and passed to organizational leaders several false defenses (pretextual reasons) in an effort to explain the low scores. Mr. ██████ repeated the pretextual reasons during the MSPB hearing. Mr. ██████ initially stated under oath that the 82's he issued me for communication and teamwork were not lower than expected. Mr. ██████ sworn statement was later impeached by the agency's Human Resources Manager and CCAS expert, Ms. ██████, when she testified that the issuance of 83s at the factor-level were "expected" scores.

Mr. ██████ admitted he did not enter any comments in the communication section of my CCAS appraisal indicating there were any issues relative to expectations concerning communication. Mr. ██████ stated under oath that he knew of a CCAS guidance document that instructed Supervisors against including negative comments in CCAS appraisals. Mr. ██████ sworn statement was later impeached by the agency's Human Resources Manager and CCAS expert, Ms. ██████, when she testified that no such document existed and that Supervisors can and should include negative comments in CCAS appraisals if warranted.

Mr. ██████ informed agency leaders he issued me the alleged low scores because I did not include enough comments in my self-assessment concerning my communication contribution. He later admitted under oath he was aware he was responsible for assessing my contributions and describing how my contributions related to the organization's mission rather than assessing how well I described each contribution in my CCAS appraisal. Mr. ██████ also admitted he was aware I was not required to enter any comments in the self-assessment section of the CCAS appraisal and that I should not be penalized for not completing that section. Mr. ██████ sworn testimony concerning his responsibilities was supported during the hearing by the enclosed appellant's Exhibit A—Supervisor's Guide for Writing CCAS Assessments (pages 2, 4, and 6). The record shows Mr. ███████, in my CCAS appraisal, actually praised my communication contribution to the agency for FY14. The record also shows the agency's leader (General ████████) agreed the communication score Mr. ████████ issued me was low and ordered Mr. ████████ to increase the score to 83, although I requested an 84 in my grievance.

Mr. ███████ admitted under oath during the MSPB hearing he did not enter any comments in the teamwork section of my CCAS appraisal indicating I had any issues relative to expectations concerning teamwork. Mr. ████████ used the same pretextual defense he used for the score he issued me for communication. His "cannot enter negative comments in CCAS appraisals" sworn testimony was impeached by Ms. ██████ and his explanation to his leaders that he issued me the scores because I did not include enough comments in my self-assessment section was impeached by Mr. ████████ sworn testimony when he admitted he was aware I did not have to enter any comments in my CCAS appraisal and that he was responsible for entering comments describing my teamwork contributions to the organization. Mr. ████████ sworn testimony concerning his aforementioned responsibilities was supported during the hearing by the appellant's Exhibit A—Supervisor's Guide for Writing CCAS Assessments (pages 2, 4, and 6).

Mr. ██████ testified under oath he received complaints about me during the FY14 rating period from Ms. ██████████ (The MSTC Project Director-PD). Mr. ████████ could not recall the nature of the complaints. Mr. ██████ admitted he did not investigate the complaints but did speak with someone about it. He testified he was not sure who he spoke with about the complaints. Mr. ██████ testified during the MSPB hearing it was his responsibility to track and resolve complaints concerning his employees. When I asked Mr. ██████ if Ms. ███████ asked him to remove me from the MSTC team Mr. ████ responded he was not sure. He gave the same response when I asked if he sought any guidance on removing me from the MSTC team. When I asked if he requested information from human resources concerning removing me from the MSTC team Mr. ████ replied "no". Mr. ████ sworn testimony concerning seeking information from human resources related to removing me from the MSTC team was incorrect. Mr. ████ sent an email message to Ms. ████ on April 11, 2013 seeking the information he claimed under oath he did not seek—information concerning removing me from the MSTC team. I purposely emphasized (above) that the message was sent to Ms. ████. I in no way alluded to any other person that might have been on the "To" line of that message. Ms. ████ did not/does not serve in a legal capacity for PEOSTRI or Mr. ████ and she received a message from Mr. ████ seeking the aforementioned information in her inbox.

I asked Mr. ████ if he remembered receiving complaints from me concerning the way Ms. ████ was treating me on the MSTC team and if some of the complaints concerned violations of law. Mr. ████ replied he could not remember. After handing Mr. ████ an email message (appellant's Exhibit B) I believed was part of the record for this case listing my complaints to him, Mr. ████ said he vaguely remembered the message. The email message documented complaints including inherent governmental functions being taken from me and given to a support contractor, the taking away of my duty to confirm delivery of Government products in accordance with an awarded contract before approving payment to the defense contractor, and Ms. ████ approving government payments to vendors before the vendor delivered the products. After reviewing my case documents for this complaint, I found that the email message was not submitted to the MSPB specifically for this case but was submitted as part of the referenced previous case—AT-1221-14-0725-W-1. The message only serves to verify that I complained about activities I believed violated law, rule, or regulation as well as evidenced an abuse of authority (the subject message is enclosed). The particular issues listed in the message, however, were submitted to the MSPB as part of this case—see Appendix C of the Appellant's Response To The Merit Systems Protection Board's Order On Jurisdiction And Proof Requirements document. When asked if he investigated the complaints Mr. ████ said he did not but he spoke to LTC ████ (the Medical Simulation Team Leader) and Mr. ████ (the Procurement Contracting Officer) about the complaints. Mr. ████ stated he did not report the complaints to his Supervisors. Mr. ████ testified during the MSPB hearing it was his responsibility to track and resolve complaints concerning his employees.

Mr. ████ conducted a serious violation during the MSPB hearing. I witnessed Mr. ████ meeting with Ms. ████, immediately after his testimony concluded and before her testimony started, and heard him discussing questions he was asked during the hearing relative to my 2014 CCAS appraisal. Mr. ████ action directly violated the Judge's direction to him ordering him not to discuss his testimony until the hearing was over. I reported what I witnessed to the Judge and was allowed to question Ms. ████ concerning the meeting. Ms. ████ corroborated my assertion.

████

Ms. ████ served as my Supervisor during the FY15 rating period and is currently my Supervisor. Ms. ████ testified she joined the Medical Simulation team (which includes the MSTC team) in October 2014 as a Chief Engineer and Supervisor and admitted she had never served as a Supervisor before. Ms. ████ testified she did not compete for the supervisory position and was added to the team as a Supervisor via a Management Directed Reassignment and added that LTC ████ told her she was the engineering Supervisor for the Medical

Simulation team. Ms. ████ sworn statement was later impeached by the agency's Human Resources Manager, Ms. ████, when she testified that Management Directed Reassignments can only be affected by the PEO—the General, not LTC ████.

I asked Ms. ████ during the MSPB hearing if she met with Mr. ████ during the time period following his testimony and before her testimony and she replied "yes". I asked Ms. ████ if they discussed anything related to the hearing and she replied "no". When I informed Ms. ████ I heard Mr. ████ mentioned my 2014 CCAS as I walked by, Ms. ████ changed her testimony and said Mr. ████ told her he was asked questions concerning the 2014 CCAS appraisal he completed for me.

Ms. ████ issued me low performance evaluation scores for two FY15 CCAS elements—communication and teamwork. After I formally accused Ms. ████, via an internal grievance, of issuing me low scores she created false defenses (gave pretextual reasons) in an effort to explain the scores (just as Mr. ████ had done a year earlier). Ms. ████ admitted she did not enter any comments in the communication and teamwork sections of my CCAS appraisal indicating there were any issues relative to expectations concerning the two factors. Ms. ████ made the same statement under oath that Mr. ████ made to explain why she did not enter any negative comments in my 2015 CCAS appraisal. Ms. ████ stated under oath that she was instructed against including negative comments in CCAS appraisals. Her testimony was later impeached by the agency's Human Resources Manager and CCAS expert, Ms. ████, when she testified that Supervisors can and should include negative comments in CCAS appraisals if warranted. Ms. ████ expert-impeached assertion matched the answer Mr. ████ gave and was likely passed-on to Ms. ████ by Mr. ████ during the conversation she had with him immediately following his testimony as part of the hearing. Additionally, Ms.

██████ did enter negative comments in my mid-year CCAS appraisal document (see page 18 of the agency's pre-hearing submission document). Lastly, Ms. ██████ also entered negative comments on my 2017 final CCAS appraisal (enclosed—see page 9) before her hearing testimony (i.e., before 13 October 2017). My 2017 CCAS appraisal was not available before 12 January 2018—after the hearing—but impeaches Ms. ██████ sworn testimony concerning entering negative comments in a CCAS appraisal. Therefore, Ms. ██████ testimony that she could not enter negative comments in CCAS appraisals is unbelievable, unreliable, and was knowingly false.

Ms. ██████ informed agency leaders she issued me the low scores because I did not include enough comments in my self-assessment concerning my communication contribution. She later admitted under oath she was aware she was responsible for assessing my contributions and describing how my contributions related to the organization's mission rather than assessing how well I described each contribution in my CCAS appraisal. Ms. ██████ also admitted she was aware I was not required to enter any comments in the self-assessment section of the CCAS appraisal and that I should not be penalized for not completing that section. Ms. ██████ sworn testimony concerning her aforementioned responsibilities was supported during the hearing by the enclosed appellant's Exhibit A—Supervisor's Guide for Writing CCAS Assessments (pages 2, 4, and 6). Although Ms. ██████ neglected to include any negative comments in the communication section of my CCAS appraisal, she created negative comments to send to agency leaders in response to my grievance. Ms. ██████ informed that she "provided input to Mr. Ingram regarding his communications skills based on my observations with his interactions with other team members, input that he did not accept." (see page 18 of the agency's pre-hearing submission document). Those comments are false and would have been included in my annual

CCAS appraisal by Ms. ███ if they were true. Also, Ms. ███ has never mentioned a specific communications issue I displayed. Ms. ███ also admitted in the same document she only used the two (2) communication contributions she included in her CCAS section when she gave me the score of 82. She did not use my other 11 key communication contributions to decide on the score.

Ms. ███ also neglected to include any negative comments in the teamwork section of my CCAS appraisal. However, she created negative comments to send to agency leaders in response to my grievance. Ms. ███ informed that I failed to promote an "environment of teamwork and cooperation" and failed to apply innovative "approaches to resolve difficult issues while working with others…." (see page 18 of the agency's pre-hearing submission document). Those comments are false and would have been included in my CCAS appraisal by Ms. ███ if they were true. Also, Ms. ███ did not mention any of those issue during the hearing. Ms. ███ also admitted in the same document she only used the two (2) teamwork contributions she included in her CCAS section when she gave me the score of 82. She did not use my other 13 key teamwork contributions to decide on the score.

Ms. ███ stated she remembered removing me from the MSTC Lead Engineer position. She said she moved me to a non-Lead Engineer position on the VHA team. When I asked Ms. ███ if I led any VHA IPTs (Integrated Teams) she replied "no". She admitted I did lead IPTs that included civilian and military personnel while serving as the MSTC Lead Engineer (verifiable by my CCAS appraisals signed by Ms. ███). Ms. ███ also admitted I led teams that were responsible for delivering products to various MSTCs around the country as well as led a team that was responsible for developing a new MSTC.

Ms. ████ admitted she gave my Lead Engineer MSTC position to Ms. ████ after she moved me to the non-Lead Engineer position on the VHA team where Ms. ██ served before being moved. Although Ms. ████ stated she did not feel I was more qualified to serve as the Lead Engineer on the MSTC team than Ms. ██, she admitted she had to return Ms. ██ to the VHA team shortly after she moved her to the MSTC team. Ms. ████ admitted I served successfully as the MSTC Lead Engineer for several years.

Ms. ████ initially told Attorney ████, in response to a hearing question, there were no Lead Engineers at PEOSTRI. However, Ms. ████ later said she was once a Lead Engineer. Ms. ████ repeatedly referred to me as the Lead Engineer in her CCAS appraisal comments (see her MID-YEAR SUPERVISORY INPUT document submitted to the MSPB in Enclosure 2 of the Appellant's Response To The Merit Systems Protection Board's Order On Jurisdiction And Proof Requirements document).

I questioned Ms. ████ concerning discriminatory treatment related to a mode of travel I was forced to take to a TDY location. I asked Ms. ████ if she refused to allow me to drive a rental car to FT Stewart for a TDY event and she responded "yes" per the travel regulations. Ms. ████ testified she told me I had to reserve and drive one of the organization's vehicles if one was available. I asked her if she refused to allow me to fly when I asked to change my mode of travel and she replied "yes". When I asked about the regulation used to refuse to allow me to fly, Ms. ████ was not able to name a regulation that supported her decision to refuse to allow me to fly to the TDY location. When I asked Ms. ████ if she later allowed another employee to drive a rental car to FT Stewart for a TDY event even though the organization had a vehicle available she replied "yes". When asked why she treated me differently Ms. ████ said the employee lied to leadership to gain approval to drive a rental car. Upon further questioning, Ms.

██████ admitted the employee had already reserved one of the organization's cars when she approved him for a rental car. Ms. ██████ had no explanation when asked why she approved the rental car when she knew the other similarly situated non-whistleblower employee had reserved an agency vehicle.

██████████

Ms. ██████████, the agency's Human Resources Manager and CCAS expert, described her position, during the hearing, as the Human Resources Supervisor. A position she said she has held at PEOSTRI since 2011. Ms. ██████ was asked by Attorney ██████ to explain CCAS and Ms. ██████ gave the educational explanation. Ms. ██████ was asked how she knew me and she responded she knew me from a previous complaint. Ms. ██████ was asked about Management Directed Reassignments. Ms. ██████ shared that Management Directed Reassignments can only be ordered by the PEO (i.e., the General Officer at PEOSTRI). This testimony directly contradicted Ms. ██████████ earlier testimony where she stated LTC ██████ used a Management Directed Reassignment and told her she was the engineering Supervisor of the Medical Simulation team.

Ms. ██████ responded "yes" when asked by the Judge if a score of 83 was an expected level score. She added the score was the same for all "maxed-out" employees. Ms. ██████ also responded "yes" when asked if additional pay is given if the employee scores higher. She added, when asked by the Judge how higher factor scores could affect compensation, that higher factor scores could change the contribution award amount.

I asked Ms. ██████ if she provided CCAS training to Supervisors and she said "yes". After hearing Ms. ██████ refer to a score of 83 as expected on multiple occasions, I asked her if the factor score of 83 on my CCAS appraisal was an expected score and she replied "yes" after explaining that the scoring range was 80 to 87. This testimony directly contradicted Mr. ███████ and Ms. ███████ earlier testimony where they both stated the 82's they issued me for communication and teamwork were not lower than expected (Ms. ███████ matching testimony came shortly after she was seen and admitted to discussing Mr. ███████ hearing testimony with him in the hallway before she entered the hearing to give her testimony). When I asked Ms. ███████ if she was aware of any policy or directive that informed Supervisors against entering negative comments in their section of the CCAS appraisal she replied "no" and added that the Supervisor should include negative comments if warranted. This testimony directly contradicted Mr. ███████ and Ms. ███████ earlier testimony where they both stated a policy existed that kept them from entering any negative comments in my CCAS appraisal (testimony expected to explain why they gave me lower than expected/usual scores for communication and teamwork without including justifying comments in the CCAS appraisal).

Protected Disclosures Contributing Factor in Personnel Actions

I believe I have supported my claim that I engaged in protected whistleblowing and that the agency took multiple punitive personnel actions against me. The personnel actions taken by the agency were taken less than a year after my disclosures (with the exception of my 2008 disclosures).

Mr. ███████ issuance of the low FY14 performance appraisal factor scores, which led to a low overall score and diminished my annual financial award, was done less than a year after I disclosed to the Procurement Contracting Officer, Military Officers, and other team members (as well as the OSC and the MSPB) that acquisition rules and laws were being violated by the PD (███████) and that Mr. ███████ was aware of the actions but did nothing to address the violations. Mr. ███ was also aware of my 2008 MSPB case. He testified he was made aware of my past MSPB case, where I named his Supervisors as wrongdoers and the MSPB ordered the agency to return me to the MSTC team and raise my annual review score, by his Supervisor Ms. ████████.

Ms. ██████ removal of my Lead Engineer position and duties as well as her issuance of the low FY15 performance appraisal factor scores, which led to a low overall score and diminished my annual financial award, was done less than a year after she became aware of my 2008 and 2014 MSPB cases where I accused her current and former supervisors of wrongdoing. Ms. ██████ testified she became aware of my past cases soon after she joined the medical simulation team after she was told to google my name. I also questioned Ms. ██████ Supervisor in my past complaint (Mr. ████████, a key person in my 2008 MSPB complaint) concerning Ms. ██████ irregular promotion to her current supervisory position—see the last page of Appendix B of the Appellant's Response To The Merit Systems Protection Board's Order On Jurisdiction And Proof Requirements document. It is likely Ms. ██████ became aware that I disclosed what I believed to be a violation of policy, rule, or law concerning her promotion without competition.

The record directly and circumstantially supports that I engaged in protected whistleblowing activity and that protected activity was a contributing factor in the agency's decision to take several personnel actions against me. Although the agency offered a few reasons for taking the actions against me, all of those reasons were contradicted by the agency's expert witness, Ms. ████████, or by Mr. ██████ and Ms. ██████ own sworn testimonies. The agency had no legitimate reason for taking the personnel actions against me and offered no evidence to support that my Supervisors took similar actions against other employees who were not whistleblowers but were otherwise similarly situated. Ms. ▢ simply responded "yes" when asked if she had taken similar actions against other employees. Ms. ▢ numerous inconsistent statements as well as her demeanor during the hearing makes her "yes" response very questionable. The agency did not establish by clear and convincing evidence that it would have taken the same personnel actions against me in the absence of my disclosures. A reasonable person would conclude that my disclosures were a contributing factor in the personnel actions my agency took against me.

Knowledge and Timing Test

The actions taken by my leaders (▢ and ▢) meet the knowledge-timing test verifying that my disclosures were a contributing factor in the unfair personnel actions they took against me. Regarding the timing prong of the knowledge-timing test, the relevant inquiry is the time between when the agency official took the action against me and when he/she had actual or constructive knowledge of my disclosure (not necessarily the date of the disclosure itself). See Caddell v. Department of Justice, 57 M.S.P.R. 508, 514 (1993). The knowledge-timing test can be satisfied where the disclosure and personnel action are 2 years apart. See Gonzalez v. Department of Transportation, 109 M.S.P.R. 250, ¶ 19 (2008).

Mr. ▢ served as my Supervisor from FY12 through FY14. He was made aware of my initial MSPB case by Ms. ▢ at some point after he became my Supervisor. Mr. ▢ also became aware in 2014 that I reported, to several PEOSTRI leaders, the OSC, and the MSPB, several actions taken by my team leaders that I believed violated acquisition rules and laws and that he knew about the illegal activities and did nothing to address the issues. Because Mr. ▢ issued me the punitively low performance evaluation scores in my FY14 CCAS appraisal in October of 2014 (less than two years after he became aware of my initial MSPB complaint where I implicated his Supervisors and a year after he became aware of my 2014 disclosures), a connection between my disclosures and the personnel actions Mr. ▮▮▮ took against me is evident and the knowledge-timing test is satisfied. Since Ms. ▮▮▮ informed Mr. ▮▮ of my earlier disclosures to the OSC and the MSPB, support also exists for constructive knowledge. Constructive knowledge may be established by demonstrating that an individual with actual knowledge of the disclosure influenced the official(s) accused of taking the retaliatory action. See Marchese v. Department of the Navy, 65 M.S.P.R. 104, 108 (1994).

Ms. ▮▮▮▮ served as my Supervisor in 2015. She was made aware of my initial MSPB case as well as my 2014 MSPB case when she was told to google my name, according to her sworn testimony. Ms. ▮▮▮ became aware that I disclosed wrongdoing by her Supervisors in my initial and 2014 MSPB complaints. She also became aware I mentioned her irregular promotion to the supervisory position she currently holds in my 2014 MSPB case. Because Ms. ▮▮▮ significantly reduced my duties when she removed me from my Lead Engineer position and issued me the punitively low performance evaluation scores in my FY15 CCAS appraisal in October of 2015 (both actions less than a year after she became aware of my initial and 2014 MSPB disclosures where I implicated her Supervisors and characterized her promotion as illegal), a connection between my disclosures and the personnel actions Ms. ▮▮▮ took against me is evident and the knowledge-timing test is satisfied.

Conclusion

I accused my Supervisors of repeatedly retaliating against me by subjecting me to prohibited personnel practices for making protected disclosures. I believe the evidence highlighted during the hearing as well as what exists in the record substantiates my retaliation claims. I clearly made protected disclosures and there is no dispute that my Supervisors were well-aware of my disclosures. The agency's witnesses testified the issuance of the low CCAS scores did negatively affect my pay (i.e., I would have received a higher bonus if issued higher scores). My removal from the Lead Engineer position the MSPB ordered I be returned to substantially reduced my duties as corroborated by Ms. ███████ testimony and my approved CCAS entries. The time between when my Supervisors became aware of my disclosures and the taking of the personnel actions against me was short—less than a year in most cases. Therefore, the knowledge-timing test is satisfied. The agency, especially my Supervisors, offered multiple reasons for taking the personnel actions against me but each given reason was proven to be false (either contradicted by the agency's expert witness or contradicted by my Supervisors' testimony). Ultimately, the agency had no legitimate reason for taking the personnel actions against me and offered no evidence to support that my Supervisors took similar actions against other employees who were not whistleblowers but were otherwise similarly situated. The agency did not meet its burden of proof. General statements unsupported by other evidence do not satisfy the clear and convincing standard. See Schnell, 114 M.S.P.R. 83, ¶ 24. Additionally, my Supervisors showed contempt for the MSPB process by discussing testimonies inappropriately and knowingly giving false testimony.

I am hopeful the MSPB will recognize I have been serving the taxpayers and Servicemembers well and did/do not deserve to be unfairly punished as I have been in the past. I also hope the MSPB will find that the agency did unlawfully retaliate against me, order the agency to return me to a Lead Engineer Position on the medical simulation team, order the agency to increase my FY14 and FY15 Teamwork and Communication scores from 82 to 84, order the agency to pay me the correct amount of back pay with interest, order the agency to provide immediate supervisory training to Ms. ██████, and order the agency to administer the appropriate required disciplinary action, in accordance with public law, for Supervisors who retaliate against whistleblowers.

_____

Harroll Ingram, Pro Se

Appendix E

CLOSING STATEMENT (2018-B)

Investigator

DOD Defense Civilian Personnel Advisory Service

Investigations and Resolutions Directorate

HARROLL INGRAM   Docket No. ARPEOSTRI17JUL02493

Complainant, Pro Se,

v.

DEPARTMENT OF THE ARMY   Date: 25 January 2018

Agency.

CLOSING STATEMENTS

Title VII of the Civil Rights Act of 1964 prohibits discrimination on the basis of race, color,

religion, sex, national origin, or retaliation for participation in EEO activity. I (the complainant),

in the subject EEO Fact Finding Conference (FFC) held on January 17, 2018, set out to show

that leaders in my organization (Program Executive Office for Simulation, Training, and

Instrumentation—PEO STRI) racially discriminated against me by neglecting to allow me to

interview for, and have an opportunity to gain, one of two NH-04 supervisory Engineer

positions. I also claimed my organization retaliated against me because of my past EEO activity.

I am an African American/Black male. None of the applicants who were allowed to interview for

the two positions was black, none of the panel members was black, and several of the panel

members (Mr. [REDACTED], Mr. [REDACTED], Mr. [REDACTED], and Mr. [REDACTED]) testified they knew my

race before evaluating my resume. Of course, a fourth panel member (My Supervisor—Ms.

[REDACTED]) knew my race as well. In this current complaint I mentioned my belief that my

organization was continuing to use their selection system to systematically screen out members

of my social group (African America/Black) as I claimed in my past EEO complaint. I also made

the same claim in the current complaint but mistakenly used the term "Disparate Treatment" after defining "Disparate Impact". My organization's EEO Manager (Mr. ███████████) denied my request to edit my complaint to properly list "Disparate Impact" and the Investigator refused to investigate the Disparate Impact defined in my complaint without approval from my organization's EEO Manager.

PEOSTRI has a long-running history of not hiring or promoting black Engineers. PEOSTRI has routinely treated non-black Engineers better than black Engineers concerning hiring and promoting. A past PEO (Major General ███████) tried to cease the nepotism and unfair hiring and promoting practices (as well as other unethical actions) he found at PEOSTRI but was not successful. Major General ███████ conducted a climate survey that revealed ethical, hiring, and promoting issues. I forwarded the presentation slides Major General ███████ used, to show his survey results to the workplace employees, to the Investigator in this current EEO action.

I believe the evidence highlighted during the FFC, as well as what exists in the record, substantiates my discrimination and retaliation claims. This submittal documents my closing statements concerning the FFC.

Knowledge of Past EEO Complaint

Several people who took part in the selection process knew or should have known about my past EEO and other complaints. As stated during my testimony, weekly PEO meetings were held wherein each Program Manager and the Human Resources Manager updated the PEO on issues within their respective program offices. I am certain my past complaints were discussed at those meetings. Several of the selection panel members were high-ranking employees in their program offices and admitted they were aware of my past complaints. The Human Resources Manager, Ms. ███████, testified she knew me as a result of my past complaints. Mr. ███████, a panel member, testified he was aware of a past EEO complaint I filed but did not know what it was about and could not remember who told him about it. Mr. ██████████, the purported Selecting Official, testified he heard of my past EEO complaint in September of 2015 when he served as the acting Deputy PEO. Mr. ██████ was named during sworn testimony given during the FFC of my past EEO complaint as the person responsible for removing me from my leadership position on a past team and replacing me with a white coworker who was promoted, shortly after, to an NH4-equivalent (GS14) position (see pages 425 and 426 of the Investigator's report—pages enclosed). The actions taken by Mr. ██████ and covered in my past EEO complaint ensured a black Engineer was not promoted to the GS-14 level. Mr. ██████ actions in the current EEO complaint also ensured a black Engineer was not promoted to the GS-14 level position and continues the Disparate Impact I complained about in my past EEO case. Ms. ██████████████, a panel member and my Supervisor, was not available for the FFC but testified in a December 2017 MSPB hearing that she became aware of my past complaints via the internet when she was told to google my name. Ms. ██████ was definitely aware of my recent involvement in an EEO complaint (i.e., aware that I gave statements to an Investigator from the Office of the Deputy Assistant Secretary of the Army, during an AR 15-6 investigation, concerning a sexual harassment claim made by one of our coworkers). My statements to that Investigator included my belief that my leaders failed to address the sexual harassment claim properly.

I believe the aforementioned key personnel on the selection team who had knowledge of my past EEO activity likely considered that past activity when they assisted Mr. ██████████, the Selection Chairperson, with creating the Go/No-Go criteria that was used as a pretext for discrimination (i.e., that was used to deny me the right to compete for an NH-4 supervisory engineering position). It is also likely that the selection team members who knew of my past EEO complaint shared that information with the other team members.

Arbitrary Go/No-Go Criteria

The job announcement created by ████ required the applicant to agree he/she would obtain a SPRDE Level-3 certification within 24 months of being granted the position being offered. Once all application packages were submitted, the Civilian Personnel Advisory Center (CPAC) reviewed and approved the qualifying application packages. The CPAC returned the qualifying application packages to Mr. ██████. Mr. ██████ after receiving the CPAC-qualified resume packages, created a separate and additional criterion that allowed him to deem me not eligible (i.e., deem my resume a "NO-GO") to compete for the two supervisory engineering positions and inform the evaluation panel not to score my resume.

Mr. █████ debriefed me on the selection actions on or about July 20, 2017 and mentioned things that would have helped me score higher to get an interview even though he knew the panel had not scored my resume. Mr. ██████ never mentioned the Go/No-Go criteria he created after receiving my application package (i.e., after receiving my resume, questionnaire with answers, SF-50, and education transcripts). Everything Mr. ██████ mentioned to me during the debrief was in the application package I submitted in response to the posted announcement. That left me wondering why my resume did not score high enough to earn me an opportunity to interview for either of the positions.

Mr. ██████, according to his testimony, received the application packages from the CPAC on December 13, 2016. However, I received a letter from the CPAC on November 25, 2016 (three weeks earlier) informing me that my package had already been forwarded to Mr. ██████. Mr.

███ testified that he and the panel members created a resume evaluation criterion that required me and the other applicants to show in our resume that we already obtained a SPRDE Level-3 certification. Mr. ███ also testified that the resume evaluation criteria were finalized (i.e., approved for use by the evaluation team) in December 2016. Mr. ███ testimony supports my claim that he, with assistance from the resume evaluation panel members, created the arbitrary Go/No-Go criteria after receiving my resume. Since the must-already-have-a-SPRDE-Level-3-certification requirement did not exist during the period the announcement was open for competition, I did not know I was required to include the SPRDE Level-3 certification in my resume to be allowed to compete for the announced position(s). Consequently, Mr.

███ did not allow my CPAC-approved application package to be included in the competition for the two supervisory engineering positions. The arbitrary and capricious certification criterion was whimsically created by ███ and violated the agency's policy that informed its leaders that selection processes must be carried out in accordance with the PEO STRI Recruitment and Selection Guide and the Uniform Guidelines on Employee Selection Procedures. The PEO STRI Recruitment and Selection Guide states that DAWIA certifications (e.g., the SPRDE Level-3 certification) will not be used as a screen out factor.

Mr. ███ testified that the certification requirement was in the initial announcement. Mr. ███ directed the Investigator to the "Key Requirements" section of the announcement where it stated Must be able to obtain and maintain a Secret security clearance; Must sign a 3-year tenure agreement upon entering this position; Acquisition Corps membership at appointment is required; and See 'Other Requirements'. Mr. ███ then asked the Investigator to see the "Other Requirements" on a following page. On the following page, the "Key Requirements" continued. The "Other Requirements" included Selective service registration for males born after

1959; File an annual financial disclosure statement; and Requires Senior Level III certification in SPRDE. Mr. ███ failed to address the fact that the "Key Requirements", including the "Other Requirements", are listed to inform the applicant of requirements that will apply once the applicant has been selected for one of the two positions—not required during the competition phase of the selection process. The SPRDE certification requirement that applies during the competition phase is listed in the questionnaire document that accompanies the announcement. The section of the questionnaire titled "Assessment Questionnaire" (page 3, number 5) requires the applicant to agree to obtain a SPRDE Level-3 certification within 24 months of being granted one of the two positions being offered. The record shows I answered, "Yes, I am able to obtain a Senior Level III certification within 24 months of employment". In fact, and as testified at the FFC, I have held a SPRDE Level III (Senior) certification for over a decade and am seeking a secondary Level-3 certification in Program Management. The majority of the review panel members, including my Supervisor, were aware I already had a SPRDE Level III certification.

Improper Selection Process

The selection process used to fill the two positions did not follow the agency's policy. The agency's policy requires the use of the Uniform Guidelines on Employee Selection Procedures as well as the PEO STRI Recruitment Selection Guide.

The PEO STRI Recruitment and Selection Guide requires the Panel Chairperson to submit the selection memorandum with supporting documentation to the Selecting Official for signature. Several panel members (███████████████████████████████████) testified they did not submit their resume evaluation notes to the Panel Chairperson to be submitted with the selection memorandum. The HR Manager refused to directly answer the question of whether or not the supporting documentation was submitted with the selection memorandum—she responded, "Now you know I would have returned it if the notes were not included." Although a couple of the panel members offered to locate and submit their notes to the Investigator after the FFC, those notes are no longer reliable since they might be edited by the time the Investigator receives them. Failure to properly retain selection records violates 29 C.F.R. §1602.14. The hiring official should maintain all interview notes and documents related to the application and selection process in a secure centralized location for two years.

The PEO STRI Recruitment and Selection Guide requires the EEO Manager to appoint a trained Observer to ensure compliance with EEO principles during all phases of the selection process and to observe selection panel deliberations/review of the resume evaluation results. The assigned Observer (█████████████) was not present during the Go/No-Go resume evaluations. Mr. ████████ testified he attended one meeting where the resume scores were discussed. He also testified he was not aware there was a separate Go/No-Go resume evaluation meeting held.

The PEO STRI Recruitment and Selection Guide states, "DAWIA certification will not be used as a screen out factor." The document adds, "Possession of certification that is not required by the qualification standard and which may be obtained after entering a position will not be used as a screen out factor." The agency's announcement for the subject positions required the applicant to agree to obtain a SPRDE Level-3 certification within 24 months of being granted the position. Mr. ████████ decision to create a Go/No-Go criterium that required me to have a SPRDE Level-3 certification listed in my resume to compete for the positions was improper and violated the agency's policy.

Mr. ████ (the purported Selecting Official) testified two NH-4 Supervisory Engineering positions became vacant and he and Mr. ██████ met to discuss plans to fill the positions. Mr. ██████ testified they "discussed who was available" and "Potentials" in engineering. Mr.

██████ changed his testimony after the agency's lawyer interrupted his testimony. It is likely Mr. ██████ and Mr. ██████ pre-selected potential replacements to fill the vacant positions I applied for and chose to find a way to screen me out to ensure the pre-selected individuals were granted the positions.

Conflicting Testimonies

██████████ testified that having the Level-3 SPRDE certification was critical to serving in either of the two positions. However, Mr. ██████ testified the Level-3 SPRDE certification Go/No-Go criterium was developed to show ability to manage. Because the Level-3 SPRDE certification is required for NH-3 non-supervisory engineering positions, having the certification does not show the holder is equipped to serve in an NH-4 position and does not show the holder is able to manage/supervise other employees. The EEOC has found that inconsistencies in statements related to selection activities give rise to suspicion and might be significant enough to warrant a finding that, more likely than not, the selection was based on improper criteria.

Mr ██████ testified he served as the Selecting Official for this selection because COL ██████ (his Supervisor and the division's Leader) had recently started working at PEO STRI. Later, Mr. ██████ testified COL ██████ was not new to PEOSTRI when the selection activities for the two positions started. Mr. ██████ testified that either Mr. ██████ or COL ██████ was the Selecting Official. Mr. ██████ also testified COL ██████ signed the final selection memorandum— normally signed by the Selecting Official. Several of the panel members indicated they were not sure who the Selecting Official was.

Mr. ██████ testified Mr. ██████ gave him the Go/No-Go criteria before the resumes arrived. However, the CPAC email message, dated November 25, 2016, informed that my resume had already been forwarded to Mr. ███████. The Go/No-Go criteria were approved in December 2016 according to Mr.████████. Therefore, the resumes arrived before the Go/No-Go criteria was approved for use and passed on to Mr. ██████ and the other panel members.

Mr. ████████ testified he did not hear any discussions about my resume. However, Mr. █████ testified the group discussed the fact that I had the SPRDE Level-3 certification but they had to exclude me because the certification was not listed in my resume—to be fair.

Ms. ██████ indicated she received the supporting documents with the selection memorandum when she said, "You know I would have returned it if the supporting documents were not included." However, not one panel member testified they submitted their evaluation notes to be included with the selection memorandum. Several of the panel members testified they did not submit their notes. Therefore, the supporting documents were not submitted to Ms. ██████ (the Human Resources Manager) as she indicated. Not submitting the evaluation notes violated the organization's and agency's policies.

Plainly Superior

I believe my credentials showed I was better suited than the selectees for either of the two filled supervisory engineering positions. If my right to be interviewed for the positions had not been violated, I would have been granted one of the two positions. My resume showed a depth, length, and breadth of experience that surpassed that of the selectees. My resume showed I did higher-grade work than the selectees earlier in my career when I served as a Radar Tracking Systems Test Engineer for the E2-C Hawkeye surveillance aircraft as well as when I served as the Lead Engineer, Test Director, and Information Assurance POC for several PEOSTRI programs.

My education exceeds the selectees' education. My Associate Degree is in Computer Networking Technology, my Bachelor's Degree is in Electronics Engineering, my Master's Degree is in Management, and my Doctorate is in Organizational Leadership—all more relevant to the engineering leadership (supervisory) positions than the Degrees held by the selectees. I earned my Doctorate Degree by studying current concepts such as transformational leadership, critical thinking, organizational theory, management philosophies, and contemporary systems management. One of the selectees does not hold a Doctorate Degree or a Master's Degree—only has a Bachelor's Degree. It is likely the announcement requirement that states the applicant must have Bachelor's Degree (no Master's or Doctorate Degree required) was designed to help that selectee gain one of the positions. A Bachelor's Degree is historically required for entry-level engineering position. Arguably, having only a Bachelor's Degree does not qualify a person to supervise/lead Engineers who hold Master's and Doctorate Degrees.

My experience was also more relevant than the selectees' experience. I served as an Engineer on far more PM TRADE (PM GCTT) projects for more time than the selectees. I had over 26 years of DOD acquisition experience which was far more than the selectees—equaled their combined DOD experience. My DAWIA certifications also outweigh those held by the selectees. While I matched the selectees' SPRDE Level-3 certification, my resume showed I have a Program Management Level-2 certification that neither of the selectees have. My resume also showed I held a Lean Six Sigma Green Belt certification. My credentials and qualifications were plainly superior when compared to the credentials and qualifications of the selectees who were given the two supervisory positions. The organization's defense of its selection process was simply a pretext for discrimination.

Conclusion

I accused my organization of racially discriminated against me by neglecting to allow me to interview for and have an opportunity to gain one of two NH-04 supervisory engineering positions because I am African American/Black and because of my past EEO activity—thereby violating Title VII of the Civil Rights Act of 1964. Several people who took part in the selection process knew about my past EEO and other complaints. The Selecting Official was named in my previous EEO action (i.e., named during the FFC sworn testimony) as the person who removed me from a leadership position and replaced me with a white male who was promoted, shortly after, to an NH4-level position. PEOSTRI has a long-running history of not hiring or promoting black Engineers. The hiring issue evidences Disparate Impact as I claimed in my past EEO complaint and defined in the current complaint. The organization's past Executive Officer (Major General ███████) admitted to the existence of nepotism and other ethical issues within the organization when he revealed the results of his research. The selection panel, after realizing my resume was forwarded by CPAC to be included in the competition, developed announcement-conflicting "Go/No-Go" criteria and used that criteria to disqualify my resume from the competition.

The selection process used to fill the two positions violated organization's and agency's policies by not follow the organization-developed Recruitment Selection Guide and not being carried out in accordance with the Uniform Guidelines on Employee Selection Procedures. The resume evaluators' notes were not included with the selection memorandum and supporting documents, the Observer was not present during the Go/No-Go resume evaluations, the Panel Chairperson used a DAWIA certification (e.g., Level-3 SPRDE certification) to screen out my resume but The PEO STRI Recruitment and Selection Guide states DAWIA certification will not be used as a screen out factor, and the Panel Chairperson and Selecting Official met to discuss which

Engineers were available to fill the two vacated NH-4 supervisory engineering positions before the announcement was drafted.

My resume showed a depth, length, and breadth of experience that surpassed that of the selectees and my education in management and leadership exceeded the selectees' education and is more relevant to serving as an engineering Supervisor. My certifications (SPRDE Level-3, Program Management Level-2, Test and Evaluation Level-2, and Lean Six Sigma Green Belt surpass those held by the selectees. Even if one feels my credentials may not be plainly superior, my credentials do seem to be arguably better suited for at least one of the positions than the selectee. The agency's explanation for removing me from the competition, based on the record, is not worthy of belief and is, more likely than not, a pretext for discrimination.

I believe the evidence highlighted during the FFC as well as what exists in the record substantiates my discrimination and retaliation claims. The Disparate Impact defined in my complaint is also supported by the testimonies and record. I am hopeful the Army/EEOC will fully consider the record and find that my Title VII Civil Rights were violated by the agency/organization. I hope the agency will take steps to correct the unlawful actions of its employees by:

- ☐ Granting me an NH-4 position to make me whole
- ☐ Providing additional training to its employees concerning position selection processes
- ☐ Choosing a lawful alternative to the currently used selection process

---

Harroll Ingram, Pro Se

Appendix F

INFORMAL BRIEF (2019)

UNITED STATES COURT OF APPEALS FOR THE FEDERAL CIRCUIT

HARROLL INGRAM   Docket No. 2018-1249

Petitioner

v.

DEPARTMENT OF THE ARMY   Date: 8 January 2019

Respondent

.

INFORMAL BRIEF OF PETITIONER

This informal brief is in response to an order issued by the MSPB (hereafter referred to as the

Board) relative to a whistleblower case (Docket number AT-1221-18-0264-W-1). The order

became final on 24 October 2018.

1. Have you ever had another case in this court? Yes; In a United States District Court? Yes; In

the Equal Employment Opportunity Commission? Yes. If so, identify each case.

- □ United States Court of Appeals for the Federal Circuit, Docket No. 2015-3110

- □ DA Docket No: ARCESAV11JAN04975   Date: 26 February 2013

- □ United States Court of Appeals for the Eleventh Circuit, Docket No. 6:16-cv-00150-
  RBD-TBS

- □ United States Court of Appeals for the Federal Circuit, Docket No. 2018-2415

2. Did the MSPB or arbitrator incorrectly decide or fail to take into account any facts? Yes. If so,

what facts?

a.) The AJ incorrectly decided the agency demonstrated by clear and convincing evidence that it

issued the appellant the reprimand for legitimate reasons. Since the reasons given by the agency were false and unsupported by the record, the appellant believes the AJ's ruling was arbitrary, capricious, unsupported by substantial evidence, and therefore an abuse of discretion.

b.) The AJ failed to take into account that the threatening statement ████████ (the appellant's second-level supervisor) alleged was made by the appellant to ██████ (the appellant's first-level supervisor) was not clearly substantiated in the record. Although ████ (████ witness during a counseling meeting between ██████ and the appellant) included the statement in an email to ████████ and ██████ claiming she heard it, ████ also stated her emailed recollection was a draft and subject to change if ████████ or ████████ wanted her to change it. ██████ also stated she might change her recollection after reading the appellant's view of what was said during the counseling meeting. Not even ████████, the three witnesses who heard the conversation from the adjoining room, or the appellant corroborated ██████ recollection. If the AJ had properly taken into account the full record concerning the alleged threatening statement, the AJ's would have questioned the validity of the statement and other allegations in the reprimand letter issued to the appellant by ████████. The AJ would not have found that the agency met the clear and convincing statute.

c.) The AJ failed to take into account the record showed the appellant informed ████████ that he had removed ████ from his email list, as ██████ ordered, prior to ████████ issuing the reprimand for insubordination. If the AJ had properly taken into account the full record, the AJ would have questioned the validity of the insubordination allegation in the reprimand letter and would not have found that the agency met the clear and convincing statute.

d.) The AJ failed to take into account that the appellant's supervisor (██████) gave the appellant permission to speak with his attorney before he complied with her order to send a follow-up union-related message to ████ from his work computer. The appellant was punished earlier for sending a union-related message from his work computer. If the AJ had properly considered the permission given by ████████, the AJ would have recognized the appellant was not insubordinate for not immediately acting on the seemingly unlawful order ██████ gave him (to send a follow-up union-related message to █████ from his work computer), would not have stated, "No supervisor is obliged to wait for an employee's outside attorney to bless her routine workplace instructions before they are obeyed….", would have questioned the validity of the insubordination allegation in █████████ reprimand letter, and would not have found that the agency met the clear and convincing statute.

e.) The AJ failed to take into account that ████████ assumed the appellant had not followed her order and █████████ based his insubordination allegation, included in his reprimand letter, on ████████ assumption. The AJ should have recognized no evidence of insubordination existed when ████████ decided to reprimand the appellant. If the AJ had properly considered that an assumption was the basis for the reprimand, the AJ would have recognized the appellant was not insubordinate, would have questioned the validity of the insubordination allegation in ████████ reprimand letter, and would not have found that the agency met the clear and convincing statute.

f.) The AJ failed to take into account that ████████ decided to reprimand the appellant before investigating whether or not the appellant had been insubordinate or had made false statements. ████████ testified he had not conducted an investigation, was purely acting on ████████ words, and told █████ his job was to protect her. If the AJ had properly considered that ████████ failed to get the appellant's version of events before deciding to punish the appellant, thereby violating the appellant's due process rights, the AJ would have recognized the reprimand was fraudulently issued and would not have found that the agency met the clear and convincing statute.

g.) The AJ failed to take into account that the record showed ███████ threatened the appellant with reprisal if he did not follow ████████ order. The threat was issued shortly before ████████ actually reprimanded the appellant for not following ████████ order. If the AJ had properly considered ████████ threat, the AJ would have recognized the reprimand was previously planned, was a pretext for retaliation, and was fraudulently issued. The AJ would not have found that the agency met the clear and convincing statute.

h.) The AJ incorrectly decided the appellant refused to acknowledge that he would comply with ████████ instructions to remove any employees from his email list who did not want to be involved in the union-starting effort. The record does not show the appellant was instructed to acknowledge he would comply with ████████ instructions to remove any employees from his personal email list. If the AJ had not incorrectly added the "refuse to acknowledge" claim, The AJ would not have used that additional claim to support his insubordination finding.

i.) The AJ failed to take into account the record showed ████████, ████████, and several other managers within the organization exchanged several preparatory email messages and took other actions leading up to the issuance of ████████ reprimanding letter. The agency's preparations carried over to the issuance of ████████ reprimand. The AJ failed to recognize that those activities infused motive (See Sullivan, 720 F.2d 1266, 1269, 1276 (Fed. Cir. 1983) and, along with ████████ desire to protect ████████, motivated ████████ to retaliate against the appellant.

j.) The AJ failed to take into account the record showed ████████ sought years-old email messages from a past leader (████████████) in an attempt to show the appellant in an unfavorable light. In her email message to ██, ██████ exclaimed she should have written the appellant up earlier. ██████ given reason for making that statement, when questioned during the hearing, was she made the statement in response to the appellant filing a MSPB complaint against her. This exchange is evidenced 113 minute and 30 seconds into ██████ hearing testimony. The AJ should have considered ██████ animus, motive, and propensity to retaliate against the appellant and found that the agency had not met the clear and convincing statute.

k.) The AJ failed to take into account the record showed ██████ held a clear bias favoring ██████ and against the appellant and made several inconsistent statements during his hearing testimony. ██████, during the hearing, blurted out that the appellant made the same threat to him that ██████ claimed the appellant made to ██████ (i.e., threatened ██████ by saying if he knew what was going to happen to him if he didn't rescind the letter, he would not give it to the appellant in the first place). This exchange is evidenced 55 minutes into ██████ hearing testimony. The AJ should have recognized ██████ blurted allegation was clearly unfounded. ██████ animus toward the appellant led him to blurt-out the false accusation during the hearing. If the AJ had properly considered ██████ unfounded hearing statement, the AJ would have recognized ██████ propensity to falsify allegations against the appellant. The AJ also would have found the reprimand to likely contain fraudulent allegations and would not have found that the agency met the clear and convincing statute.

l.) The AJ failed to properly consider that Ms. ██████, a potential witness requested by the appellant, was the appellant's coworker, was a whistleblower, and had filed a retaliation complaint against ██████. The AJ should not have disallowed ██████ as a witness and should have found ██████ testimony to be relevant to the appellant's case.

3. Did the MSPB or arbitrator apply the wrong law? Yes. If so, what law should be applied?

a.) The AJ misapplied the law when he stated the appellant was obliged to comply with an order that is perceived to violate a law, rule, or regulation. The U.S. Court of Appeals for the Federal Circuit affirmed that the "right to disobey" provision of the whistleblower law (5 U.S.C. § 2302(b)(9)(D)) applies when an employee is asked to violate federal law. The appellant clearly informed his supervisor he feared repercussions if he used his federal work computer to send another union-interest email message to ████, his federal coworker, as ordered. Because the appellant had a reasonable belief the action ██████ ordered violated federal law, he had a right to question that order. See Rainey v. MSPB , No. 15-3234 (Fed. Cir. 2016). The record shows

██████ told ████████ the appellant did not trust them and felt he was being setup. The appellant had a reason to believe his supervisors might direct him to take an action that could violate law. The agency had a history of ordering the appellant to take actions that violated law. See Ingram v. Dept. of the Army, 2011 MSPB 71.

b.) The AJ misapplied the law when he wrote, "No supervisor is obliged to wait for an employee's outside attorney to bless her routine workplace instructions before they are obeyed...." ████████ instruction to the appellant was not of the routine workplace type since her instruction concerned the appellant's use of his personal email list and message sent using his personal email account from his personal computer from his home during non-working hours (on July 4th). The AJ's finding also lacked consideration for the fact that the appellant's email message resulted in no detrimental effect on job performance. The organization's employees read various email messages at work on a daily basis without detriment to their job performance. The Board has held that 5 U.S.C. § 2302(b)(10) prohibits agency officials from penalizing their employees for conduct that has no adverse impact on their job performance or on the ability of others to perform their jobs. See Allred v. Department of Health and Human Services, 23

M.S.P.R. 478, 479 (1984). In Allred, the Judge issued a ruling in favor of the appellant based on a finding that no nexus existed between the appellant's off-duty conduct and his official duties.

c.) The AJ misapplied the law concerning the appellant's right to due process when he failed to properly consider ████████ testimony wherein he stated his mind was made up before he met with the appellant to discuss possibly rescinding ████████ letter. The statement is evidenced 46 minutes and 40 seconds into ████████ hearing testimony. The court ruled, in Armstrong v. Manzo, 380 U.S. 545, 552 (1965), the litigant should be given an opportunity to be heard at a meaningful time and in a meaningful manner. See also Little v. Streater, 452 U.S. 1, 5-6 (1981). ████████ did not give the appellant an opportunity to be heard before issuing the reprimand. The AJ should have recognized ████████ bias against the appellant, considered the court's ruling concerning the appellant's right to due process, found the insubordination allegations against the appellant were not substantially supported by the record, and ruled the agency had not clearly and convincingly shown they would have taken the same actions against the appellant regardless of the appellant's past disclosures.

d.) The AJ also should have recognized ████████ and ████████ abuse of authority. An abuse of authority is an "arbitrary or capricious exercise of power by a federal official or employee" that harms the rights of any person or that personally benefits the official/employee or their preferred associates. See Elkassir v. Gen. Servs. Admin., 257 F. App'x 326, 329 (Fed. Cir. Dec. 10, 2007).

e.) The AJ misapplied the law concerning only giving weight to statements that are supported by the record. The AJ's favorable agency decision relied, in part, on the agency's unfounded allegation that multiple people sought assistance from ████ in getting removed from the appellant's email list. The AJ included footnote information (Footnote #10) in his initial decision showing his reliance on the allegation. The record shows only one person (▮▮▮) sought

▮▮▮▮▮▮ assistance. The court found in Weaver v. Department of the Navy that weight should

only be given to statements that are supported by the record. See Weaver v. Department of the

Navy, 2 M.S.P.R. 129, 133 (1980).

f.) The AJ misapplied the law concerning ignoring evidence offered by the appellant. Although

the appellant claimed only one person sought Johnson's assistance in being removed from his

email, provided evidence to the AJ that no one else sought ▮▮▮▮▮assistance, and attempted

to support his position via hearing testimony from ▮▮▮▮▮, the AJ stopped the appellant's

questioning of ▮▮▮▮▮ on the issue and subsequently relied on ▮▮▮▮▮ unfounded claim

that multiple coworkers sought ▮▮▮▮▮ assistance in being removed from the appellant's list.

The AJ included a footnote (Footnote #10) in his initial decision showing his reliance on

▮▮▮▮▮ allegation. The Federal Circuit reversed the Board's decision in Whitmore, where

the appellant was accused of insubordination, because the Board excluded or ignored evidence

offered by the appellant. See Whitmore v. Department of Labor, 680 F .3d 1353, 1366 (Fed. Cir.

2012).

g.) The AJ misapplied the law concerning only giving weight to statements that are supported by

the record. The AJ's favorable agency decision relied on the AJ's arbitrary finding that the

appellant "brazenly challenges a supervisor's legitimate authority by seeking legal advice invited

discipline for insubordination." The record shows the appellant's supervisor (▮▮▮▮) told the

appellant she did not mind if he sought a legal opinion. The AJ abused his discretion when he

substituted his view in place of the actual record. According to the law, weight should only be

given to statements that are supported by the record. See Weaver, 2 M.S.P.R. 129, 133(1980).

h.) The AJ, again, misapplied the law concerning only giving weight to statements that are supported by the record when he found that the appellant's behavior concerning false statements was inappropriate in the manner described within ███████ reprimand. The record shows every false statement allegation ███████ included in his reprimand was incorrect and unsupported by the evidence (i.e., the appellant did not make any false statements).

i.) The AJ misapplied the law concerning only giving weight to statements that are supported by the record. See Weaver, 2 M.S.P.R. 129, 133 (1980). The AJ stated the appellant's second message, on June 13, 2017, to prospective union members concerning his network access being suspended was sent to the entire workforce. However, the record shows the appellant sent the message to about 40 (4% of the workforce) of his coworkers who all agreed to be part of the union (i.e., agreed to receive union-development information via email). The AJ's misreading of the record is significant since the AJ characterized the appellant's message as giving the appearance to, "…hundreds of recipients within PEO STRI employees that the agency stripped the appellant of network access…." The appellant believes the AJ's reliance on the false interpretation of the record played a significant part in the AJ's decision to deny the appellant relief. The AJ wrote, immediately after the inaccurate statements noted above, "As noted above, the strength of the agency's evidence and reasons for its concern about the content of the June 13 email are deemed solid under Carr v. Social Security Administration, 185 F.3d 1318, 1323 (Fed. Cir. 1999)."

j.) The AJ misapplied the law concerning determining whether or not an agency showed by clear and convincing evidence it would have taken the same personnel action in the absence of whistleblowing (i.e., The Carr Factors). The Board as well as the Court of Appeals for the Federal Circuit have ruled that determining whether an agency showed by clear and convincing evidence it would have taken the same personnel action in the absence of whistleblowing requires the AJ to consider whether the agency had legitimate reasons for the personnel action, the existence and strength of any motive to retaliate on the part of agency officials who were involved in the decision, and any evidence that the agency takes similar actions against employees who are not whistleblowers but who are otherwise similarly situated. See Parikh v. Department of Veterans Affairs, 116 M.S.P.R. 197, ¶ 36 (2011); Schnell v. Department of the Army, 114 M.S.P.R. 83, ¶ 23 (2010); and Carr v. Social Security Administration, 185 F.3d 13 18, 1323 (Fed. Cir. 1999). The AJ neglected to fully apply the Carr factors. The following list outlines the AJ's neglected Carr considerations.

- The record shows the agency failed to submit any supporting evidence showing legitimate reasons for the personnel actions. See 3 a-h above and the petitioner's submittal to this court titled PETITION TO THE UNITED STATES COURT OF APPEALS FOR A FEDERAL CIRCUIT REVIEW OF A MERIT SYSTEMS PROTECTION BOARD'S FINAL DECISION dated 20 November 2018.

- The record shows strong motives to retaliate on the part of agency officials who were involved in the reprimand issuance decision.

- The record shows ███████ was aware of the appellant's past complaints concerning unfairly low performance review scores she issued the appellant. ██████ was also aware of the appellant's disclosures to the OSC and MSPB that she was given her current supervisory position without competition even though she had no supervisory experience. The record shows ██████ told ████████, via email—referring to the appellant, the "supervisor to employee trust is nonexistent here and will be difficult to overcome." ██████ verified the message in her hearing testimony. ████ played a dominant role in the development and issuance of ███████ reprimand to the appellant. The court ruled in Sullivan v. Department of the Navy, 720 F.2d 1266, 1269, 1276 (Fed. Cir. 1983) that despite the official recusing himself, "his dominant role in the case throughout the proceedings" infused the action with the improper motive, thereby rendering the personnel action unsustainable. ████ dominant role in ███████ actions should have been considered by the AJ and should have led to a finding of motive and a finding that the clear and convincing statute was not met by the agency.

- ████ showed animus toward the appellant when he indicated to ████ he would not rescind the reprimanding letter she issued the appellant. The record shows ████ made that statement a week before he met with the appellant to hear the appellant's side of the story concerning what led to the issuance of ████ letter. ████ also told ████ by email his goal was to protect her. That email message was submitted as Enclosure 1 of the Appellant's Response to the Agency's Discovery Delivery Response. ████ was motivated to retaliate against the appellant by his desire to protect ████.

- ████ testified he neglected (willfully) to investigate the reasons the appellant listed for seeking assistance in rescinding the reprimanding letter ████ issued him. The AJ should have recognized ████ action as contributory evidence of a motive to retaliate against the appellant. An agency's failure to conduct a sufficient investigation before bringing an action might indicate an improper motive. See Social Security Administration v. Carr, 78

M.S.P.R. 313, 335 (1998); aff'd, 185 F.3d 1318 (Fed. Cir. 1999). ███████ was also motivated to retaliate against the appellant, as claimed in the appellant's Closing Statements submittal to the AJ, because the appellant gave statements to an Army Investigator revealing the improper actions of the medical simulation team's leadership concerning responding to a sexual harassment complaint filed by the appellant's coworker (Ms. ██████████████). At the time, Mr. ████████ served as the Deputy Project Manager (the highest-ranking civilian leader) on the medical simulation team. The appellant's statements to the Investigator reflected poorly on ████████ leadership since the appellant and ████████ were civilians and ████████ was their leader.

☐ The record shows no evidence was presented showing the agency took similar actions against employees who were not whistleblowers but who were otherwise similarly situated to the appellant. The AJ misapplied the law concerning Carr when he wrote, "I further note that the appellant identified no similarly-situated comparator employee who committed similar conduct but was treated more favorably." Under Carr, the agency (not the appellant) is required to offer evidence that it takes similar actions against employees who are not whistleblowers but who are otherwise similarly situated. The court advised in Whitmore that congress indicated the agency controls all the cards including the records that could document whether similar personnel actions have been taken in other cases. See Whitmore v. Department of Labor, 680 F .3d 1353, 1366 (Fed. Cir. 2012). The AJ in the current case should have considered that the agency did not offer any evidence it takes similar actions against employees who are not whistleblowers but who are otherwise similarly situated to the appellant and should have decided the agency failed to clearly and convincingly prove it would have taken the same actions against the appellant minus the appellant's disclosures. See Carter v. Department of the Interior, AT-1221-13-2153-W-1 (December 3, 2014). In Carter, the Board decided the agency official's strong motive to retaliate and the lack of evidence that similarly situated employees were treated the same supported the finding that the agency had not met the clear and convincing statute.

k.) The AJ misapplied the law concerning witness credibility. The AJ failed to adequately apply the Board's decision in Hillen v. Dept. of the Army which lays out the factors the AJ must weigh in considering different testimony from different witnesses. The Board decided the AJ must (1) Identify the factual question in dispute, (2) Summarize all of the evidence on each disputed question of fact, (3) State which version the AJ believes, and (4) Explain in detail why the chosen version was more credible than the other version(s). See Hillen v. Dept. of the Army, 35 MSPR 453 (1987). Misapplying Hillen, the AJ found ███████ testimony to be highly credible and well supported by the record. However, ███████ changed her testimony often, as listed below, during the hearing.

- ███████ repeatedly changed her testimony concerning whether or not the appellant told her he was not going to follow her directions. Initially, ███████ said the appellant did not tell her he was not going to follow her direction to remove Ms. ███████ from his list. ███████ added that she did not know if the appellant actually sent Ms. ███████ a message informing her he removed her from his list. Then ███████ said the appellant told her he was not going to follow her direction. Later, ███████ returned to her original answer that the appellant did not refuse to follow her order. ███████ agreed she assumed the appellant was not going to follow her directions. This exchange is evidenced 18 minutes and 45 seconds into ██████ hearing testimony.

- ☐ ██████ testified she was not aware of the details concerning the appellant's MSPB case and was scheduled to speak with the organization's attorney on 14 August 2017 to gain more information. Later, ██████ testified she told COL ███ on 17 July, 2017 the appellant's MSPB case concerned an abuse of power complaint against her.

The AJ should have considered the numerous changes in ███████ testimony and should not have found ██████ to be a credible witness.

l.) The AJ misapplied the law concerning witness credibility. See Hillen v. Dept. of the Army, 35 MSPR 453 (1987). The AJ found ████████ testimony to be highly credible and well supported by the record. However, ████████ testimony was often, as listed below, impeached by the record, ██████, and himself.

- ☐ ███████ testified Ms. ██████ never told him the appellant felt the letter she issued him was done in retaliation for his past complaints. However, once faced with the email message at the hearing discrediting his statement, ████████ agreed ██████ had told him.

- ☐ ████████ testified Ms. ██████ contacted him and ██████ saying "He did it again."—referring to the appellant sending her another email message. ███████ claim had never been mentioned before and was not supported by the record. ██████ later said he did not know if there were two messages sent by Ms. ██████. He added that ██████ gave him that information and he had no reason to doubt her. Since the AJ was aware

██████ testified ████ only informed her once—not multiple times as ████████ claimed, the AJ should not have deemed █████████ testimony as credible.

- ████████ testified there were several people who were told to write the appellant and copy █████ when they ask to be removed from the appellant's email list. █████████ also claimed in his reprimand that numerous people sought assistance from ██████ in getting removed from the appellant's list. However, the record does not support that anyone other than █████ asked █████████ for assistance. ████████ testified she told █████████ she did not know of anyone else—other than Ms. ██████. This exchange is evidenced 49 minutes into █████████ testimony.

- ████████ testified he did not know of anyone who had to interact with the MSPB. However, the record supports that he knew his employee (████████) had interacted with the MSPB in a previous case concerning the appellant. █████████verified █████████ knowledge 120 minutes and 7 seconds into her testimony.

- ████████ testified he responded to the appellant immediately to correct his threat that reprisal would occur. █████████ then changed his testimony and said he made his correction within the hour. Finally, █████████ switched his time to close to within the hour. This exchange is evidenced 10 minutes and 30 seconds into █████████ testimony.

- ████████ testified he knew the appellant had not removed ██████ from his list. When asked how he knew, █████████ stated █████ contacted him and ████████ and said "he did it again". █████████ stated █████ asked for ████████ assistance more than once and added there were several people who asked several times to be removed. █████████ sworn statements did not match the record and opposed █████████ testimony that ██████ only contacted her once.

- ▢ ███████ testified, initially, he did not know what ██████ meant when she mentioned the appellant's history but soon after changed his testimony and stated he knew the history concerned an MSPB complaint.

- ▢ ███████ testified he saw an email message wherein the appellant wrote the threatening statement to ██████ (i.e., wrote if she knew what was going to happen to her if she didn't rescind the letter, she would not give it to the appellant in the first place). ███████ added he was sure the appellant wrote the statement. The email message ███████ claims he saw does not exist and is not part of the record. The AJ should have viewed ██████ claim as harmful to his credibility.

- ▢ The AJ should have noted ███████ hearing demeanor where he was very evasive and argumentative. ███████ was on the verge of being revealed to have given false testimony when the agency's attorney objected, the AJ overruled, and ███████ questioned the AJ and informed of what he felt his testimony would/should cover. This exchange is evidenced 21 minutes and 15 seconds into ███████ testimony.

m.) The AJ wrongly applied the law when he found that the appellant's 2017 Board appeal (AT-1221-17-0498-W-1), where Ms. ██████ was named as a retaliator, was not properly before the Board. The AJ referred to the 2017 appeal as "Disclosure 4" in his initial decision. The appellant responded to the AJ in writing that he added the 2017 Board appeal prior to the OSC's issuance of its final decision. See the Appellant's Reconsideration Request Concerning The Merit Systems Protection Board's Decision To Dismiss The Appellant's 4th Disclosure document. The record shows the OSC accepted the appellant's addition but decided against changing its preliminary decision. The appellant also informed in his Closing Statements submission that his 2017 disclosure was made in accordance with 5 U.S.C. § 2302(b)(8) and was submitted to the

OSC with a sufficient basis to pursue an investigation. The appellant also sent proof of his OSC submittal to the AJ by sending his response to the OSC's initial denial letter. The Board ruled in favor of an appellant in a 2008 case by finding that the key to determining whether an appellant has satisfied the exhaustion requirement in an IRA appeal is whether he provided the OSC with a sufficient basis to pursue an investigation. See Pasley v. Department of the Treasury, 109 M.S.P.R. 105, (2008) and Ward v. Merit Systems Protection Board, 981 F.2d 521, 526 (Fed. Cir. 1992). If the AJ had adhered to the established Board decision in Pasley and not decided the appellant's 2017 disclosures were not administratively exhausted, the disclosures in that case would have clearly evidenced a motive to retaliate against the appellant, would have resulted in an inability to reasonably decide the clear and convincing statute had been met by the agency, and would have resulted in an order of relief for the appellant.

n.) The AJ misapplied the law concerning allowing relevant witnesses to testify at the hearing. The AJ disallowed three witnesses (two key witnesses) the appellant requested to prove the stated reasons for ▮▮▮▮▮ issuing the appellant the reprimand letter were mere pretexts for punishing the appellant for his past whistleblowing disclosures. The disallowed witnesses submitted written statements confirming the appellant's claim that he did not make the threatening statement to ▮▮▮ as alleged by ▮▮▮▮ via email and ▮▮▮▮ in the reprimand he issued to the appellant. The AJ disallowed ▮▮▮ as one of the appellant's witnesses. One of the disallowed key witnesses, Ms. ▮▮▮▮▮▮▮▮, was the coworker who filed the sexual harassment complaint that the appellant's team leaders, including ▮▮▮▮▮, failed to properly address. One of the appellant's disclosures concerned his testimony to the investigator who was assigned to finalize a report concerning ▮▮▮▮ complaint. As a result of the sexual harassment complaint, ▮▮▮▮ was issued a caution letter by ▮▮▮▮

on the same day ███████ issued the reprimand to the appellant. Although the AJ was aware ████████ filed a retaliation complaint against ███████ (verified by ████████ testimony), the AJ refused to allow ███████ as one of the appellant's witnesses. The AJ's abuse of discretion with respect to witnesses caused substantial harm to the appellant's ability to present a complete whistleblower defense under Carr. See Whitmore v. Department of Labor, 680 F.3d at 1368 as well as Umshler v. Department of the Interior, 945 F.2d 417 (Fed. Cir. 1991).

4. Did the MSPB or arbitrator fail to consider important grounds for relief? Yes. If so, what grounds?

a.) The AJ should have considered that the agency failed to provide any evidence that the agency took similar actions against employees who were not whistleblowers but who were otherwise similarly situated to the appellant. The failure should have resulted in a ruling that the clear and convincing statute had not been met and the appellant should have been granted relief.

b.) The AJ should have considered that the appellant informed his leaders he feared negative repercussions if he used his federal work computer to send another union-interest email message to ████—his federal coworker. The record shows ███████ told ████████ the appellant did not trust them and felt he was being setup. The AJ was aware the agency had a history of ordering the appellant to take actions that violated law. See Ingram v. Dept. of the Army, 2011 MSPB 71. Therefore, the appellant had a reason to believe his supervisor's order that he send the union-related message might result in a violation. The appellant was within his right to question the order, was not insubordinate, and should not have been punished by the agency. See Rainey v. MSPB , No. 15-3234 (Fed. Cir. 2016). Also, President Trump, on June 14, 2017, signed into law the "Follow the Rules Act" (H.R. 657) which prohibits taking a personnel action against a Federal employee for refusing to obey an order that would violate a rule or regulation.

c.) The AJ should have considered that ███████, with assistance from other PEO STRI leaders, knowingly decided to issue the appellant the reprimand before gathering all of the facts (i.e., without gathering the appellant's version of events). Because the other leaders at the agency who helped ████████ were aware of the appellant's past disclosures and the agency did not provide evidence they had taken similar actions against employees who were not whistleblowers but who were otherwise similarly situated to the appellant, the AJ should not have decided the agency met the Carr factors and proved clearly and convincingly they would have taken the same actions against the appellant if he had not made the past disclosures.

d.) The AJ should have considered that █████ written statement, used by ███████ to justify the reprimand he issued the appellant, was unreliable based on her qualifying comments informing the statement was in draft form and she would change the statement if ████████ wanted her to do so. Also, recognizing that ██████ account did not match ████████ should have signaled to the AJ that the insubordination allegation in the reprimand, based on ██████ statement, was not substantiated. Additionally, if the AJ had considered there was no evidence the appellant's statements were false as alleged by ████████ in his reprimand, the AJ would have deemed the reprimand as unwarranted and likely a pretext for retaliation—as was threatened by ████████ shortly before he issued the appellant the reprimand.

e.) The AJ should have considered the inaccurate testimony ███████ gave on several occasions during the hearing (e.g., claimed several people sought assistance from ███████ in being removed from the appellant's email list, the appellant threatened him with the same statement █████ claimed the appellant made to ██████, and Ms. █████ contacted him and ██████ saying "He did it again."—referring to the appellant sending █████ another message). ████████ hearing demeanor and credibility issues should have led the AJ to decide ████████ words were not reliable as required per the ruling in Hillen v. Dept. of the Army, 35 MSPR 453 (1987).

████████ demeanor and credibility issues should have led the AJ to find the agency had not

met the clear and convincing statute and should have led the AJ to grant the appellant relief.

f.) The AJ should have considered that allowing the appellant's witnesses (████████████,

████████, and ████████) to testify would have led to proof that ██████ and

██████ made false statements under oath and that █████ written allegation against the

appellant, indicating he threatened██████, was false. The AJ's consideration of the testimonies

████████, ██████, and ██████ would have given would have led the AJ to grant the appellant

relief.

g.) The AJ should have considered that ████████████ filed a retaliation complaint against

████████ and had relevant testimony that could have discredited ████████ sworn testimony.

The consideration would have resulted in recognition that ████████ had a propensity to retaliate

against whistleblowers (matching his reprisal threat to the appellant) and would have led the AJ

to grant the appellant relief.

h.) The AJ should have considered the motives to retaliate on the part of the agency officials who

ordered the reprimand actions as well as the other agency officials who influenced the decision.

The existence of motives should have led the AJ to find that the Carr requirements were not met

and the appellant deserved relief.

i.) The AJ should have considered the appellant's "4th disclosure" (i.e., should not have decided

the appellant's 2017 disclosures were not administratively exhausted). The disclosures in that

MSPB case would have clearly evidenced motives to retaliate against the appellant, would have

resulted in an inability to rule the clear and convincing statute was met by the agency, and would

have resulted in an order of relief for the appellant.

5. Are there other reasons why the MSPB's or arbitrator's decision was wrong? Yes. If so, what reason?

a.) The Board's decision that the agency showed by clear and convincing evidence it would have taken the same personnel actions against the appellant in the absence of the appellant's protected disclosures was wrong because:

- The AJ failed to evaluate all of the pertinent evidence—including evidence that detracted from the agency's allegations. The Federal Circuit has instructed the Board to evaluate all the pertinent evidence in determining whether an element of a claim or defense has been proven adequately. See Whitmore v. Department of Labor, 680 F.3d at 1368. The Board has held that a proper analysis of the clear and convincing evidence issue requires that all of the evidence be weighed together - both the evidence that supports the agency's case and the evidence that detracts from it. See Shibuya v. Department of Agriculture, 119 M.S.P.R. 537, 37 (2013) (citing Whitmore, 630 F.3d at 1368). If the AJ had fully considered the evidence that supported the appellant's positions (he removed ▮▮▮ from his list, he did not make the threatening statement ▮▮▮ and ▮▮▮ alleged, he did not make any false statements, and his leaders have retaliated against whistleblowers in the past), the Board could not have found that the clear and convincing statute was met by the agency.

- The personnel actions were unwarranted and were mere pretext for retaliation. ▮▮▮ and ▮▮▮ testified the letters they issued the appellant were issued based on their assumption he was insubordinate. Neither ▮▮▮ or ▮▮▮ had a viable reason to assume the appellant would not remove ▮▮▮ from his email list and neither had a legal right to punish the appellant based on their unreasonable assumptions. Because the agency provided no evidence it took similar actions, based on assumptions, against other employees who were not whistleblowers, the clear and convincing statute was not met by the agency and, therefore, the appellant should have been granted relief.

c.) The Board's finding that the appellant was insubordinate because he did not immediately follow ███████ order was wrong because the AJ noted he was aware the appellant feared he might be punished for sending a second union message from his work computer as directed by ████████. The AJ acknowledged at the bottom of page 15 of his initial decision that Ms. ██████ gave the appellant permission to speak with his attorney first. Therefore, asking for and receiving permission to speak with his attorney does not constitute insubordination. The AJ was also aware the appellant expressed that he believed the action might violate some law, rule, or regulation and was aware the appellant had been punished by the agency earlier for sending a union-related message. The AJ was aware the record showed ██████ informed ████████, before ██████████ issued his reprimand letter to the appellant, the appellant believed they were setting him up. In fact, ██████████ email message to the appellant, included in Appendix B of the appellant's initial IRA letter to the Board dated February 5, 2018, informed the appellant that reprisal would happen if the appellant failed to follow ████████ direction. The evidence more than suffices to provide a layman (like the appellant) with a reasonable belief that the appellant believed following ██████ direction could violate a law, rule, or regulation. Past court rulings support the position. See Stiles v. Department of Homeland Security, 116 M.S.P.R. 263, ¶ 17 (2011). Minus sufficient proof of insubordination, the AJ should not have found the clear and convincing statute was met by the agency and, therefore, the AJ should have granted the appellant relief.

d.) The AJ failed to give weight to the fact that the appellant was issued the reprimanding caution letter from ██████, which was the basis of the reprimand ████████ issued to the appellant, for a non-existent violation (i.e., for not immediately removing ███ from his personal email list he used to send a personal message from his home during non-work hours). Since the message the appellant sent had no adverse impact on job performance and the Board has held that 5 U.S.C. § 2302(b)(10) prohibits agency officials from penalizing their employees for conduct that has no adverse impact on their job performance or on the ability of others to perform their jobs (See Allred v. Department of Health and Human Services, 23 M.S.P.R. 478, 479 (1984)), the agency had no legitimate reason for issuing the appellant ███ letter, ███ letter, or the abnormally low performance review score based on the two aforementioned reprimanding letters. The AJ should have recognized the agency's professed basis for the reprimanding letters was a pretext for retaliation in accordance with a past ruling of the 6th Circuit court. See Tingle v. Arbors at Hilliard, 692 F.3d 523 (6th Cir. 2012). The Board's decision not to grant the appellant relief was wrong.

e.) The AJ prematurely and arbitrarily decided against the appellant when he asked the appellant, near the end of the hearing, "where is the smoking gun" and added that unless the appellant can show that management's motives were not what they said they were his "case is dead". This exchange is evidenced 79 minutes and 30 seconds into ███ testimony. Because the record included evidence against the motives claimed by the agency and motives are not the sole focus for proving whistleblower retaliation cases, the AJ's statement during the hearing was an abuse of his digression and arguably evidenced a predisposition to rule in favor of the agency.

6. What action do you want the court to take in this case?

The petitioner requests that the court review the submitted information/evidence, reverse the Board's decision, and order the proper corrective actions that will leave the petitioner in the environment he would have experienced had the unlawful retaliation not occurred. The Board's determination can be overruled if it is found that testimony by the witnesses is discredited by undisputed evidence or by fact. See Riley E. Jackson, Petitioner, v. Veterans Administration, Respondent, 768 F.2d 1325 (Fed. Cir. 1985) citing Seger v. United States, 469 F.2d 292, 309, 199 Ct. Cl. 766 (1972). Evidence in the record which, when taken alone, may amount to substantial evidence and therefore support the Board's decision will often be insufficient when the trial examiner has, on the basis of the witnesses' demeanor, made credibility determinations contrary to the Board's position. See Penasquitos Village, Inc. v. N.L.R.B 565 F.2d 1074 (9th Cir. 1977). The dearth of evidence supporting the clear and convincing statute relative to the actions the agency took against the appellant supports a reversal of the Board's decision.

Because the Board's determination failed to account for contrary evidence, the findings should be deemed arbitrary and not in accordance with law. The Federal Circuit has reached similar conclusions where less-than-substantial evidence existed to support the Board's conclusion that the agency proved by clear and convincing evidence that it would have taken the same actions against the appellant had he not made the protected disclosures. See Miller v. Dep't of Justice, No. 2015-3149 (Fed. Cir. Dec. 2, 2016). The petitioner, therefore, requests that the court order the following corrective actions:

Appendix G

EEOC STATEMENT (2019)

Positional Statement

The following is my positional statement concerning the agency's actions toward me that led to my formal EEO retaliation complaint. The agency's actions evidence a clear violation of Title VII of the Civil Rights Act of 1964, the Rehabilitation Act, and the Army's Privacy Program.

Ms. ██████████, an agency Attorney, sent a copy of my unredacted complaint, showing my social security number, home address, telephone numbers, a characterization of my mental state, and charges I leveled against several PEO STRI leaders, to nine (9) people (████████████████ ██ ████████████████████) in June of 2012. ████████ claimed she released my unredacted complaint as a notification to potential witnesses in the EEO investigation. However, ████████ released my complaint to several people who were not witnesses in the EEO process and had nothing to do with the complaint. She also released the complaint to several people who were actually named as harassers in the complaint. One person, ██████, was a non-government; contracted employee and could arguably be considered the public. Ms. ████████ is a seasoned Attorney formerly of the Judge Advocate General's Corps (JAG) where she specialized in military law. Ms. ████████ knew sharing the unredacted disclosure without my consent was prohibited by law. However, she disclosed my complaint anyway thereby showing a reckless disregard for the prohibition and my professional reputation.

Ms. ████, an agency Engineer, uploaded my unredacted 2011 EEO complaint to the organization's open database that was accessible by every PEO STRI employee and contractor (roughly 1,000 people) in December of 2017. ████████ named the folder holding my unredacted complaint and several other legal documents related to my complaint "Harroll Ingram Litigation" making the folder attractive to people who scanned the database. Although ██████ was informed that she uploaded my unredacted complaint to the agency's common drive, she did not immediately remove the folder from the drive. The agency's information technology folks had to remove my confidential information nine (9) days after I asked for the information to be removed.

I implicated Ms. ███ in my 2011 EEO complaint. Therefore, it is possible Ms. ███ willfully uploaded my unredacted complaint to a location on the drive where my coworkers could see it in retaliation for my 2011 EEO complaint. In addition to facilitating an increase of my emotional distress, Ms. ███ action was likely coercive and led a would-be charging party to abandon his or her EEO claim.

Ms. ███ played a major role in the activities that led up to my 2011 EEO complaint where I alleged the creation and sustainment of a hostile working environment. Although I and the EEOC were not aware of ███ leading role in the original complaint due to the unavailability of her relevant personal email messages, those messages were later obtained via the discovery process of the subsequent federal lawsuit that resulted from the 2011 complaint. Those messages, also provided as evidence in this complaint, show ███ was the instigator and sustainer of the actions to have me removed from my leadership positions (Lead Engineer and Test Director) on the Bradley team. ███ also coached ███, a support contractor at the time, in making particular statements to my first and second level supervisors (███████████) and the team leader's (███████) first and second level supervisors (███████████) that led to my removal from the team. ███ acted in secret without the knowledge of the team leader when she took several actions to have me removed from the team. She asked ███, in one of her email messages, not to tell ███████ of her actions. ███ email messages also show she referred to me in derogatory terms (e.g., an ass) in her direct email messages to ███. ███

informed, in one message, that she feared I might file a grievance against her. She apparently knew she was conducting herself inappropriately.

██ and ██ often disrespected my leadership positions on the team. However, the team leader (████) testified I was the Lead Engineer and Test Director for the Bradley BATS and Bradley COFT-E projects. Despite ████ negative comments concerning my abilities, he informed ████, in one message, that the Bradley BATS was in the best shape ever. There was no justification for the harassing and disrespectful actions ████ and ████ took against me.

One email message that was sent after I was removed from the team shows ████ supporting ████ freestyle rap song wherein, he denigrated me and made false statements concerning my professional credibility. ████ actions evidence animus and a willful motive to retaliate against me. The evidence also shows a connection between my protected activity and ████/the agency's adverse actions against me. All of the aforementioned email messages were submitted as evidence in the subject complaint.

Mr. ████ served as a selection panel member concerning a job promotion I applied for. My resume was unfairly thrown-out of that competition by the agency—even after the Civilian Personnel Advisory Center (CPAC) approved my resume for the competition. The unfair action is the subject of an ongoing EEO complaint (ARPEOSTRI17JUL02493). Mr. ████ admitted under oath during the fact-finding conference that he was aware of my 2011 EEO complaint. I submitted that transcript section as evidence in the current complaint. I believe Mr. ████ and/or the other selection board members viewed the unredacted 2011 complaint released by Ms. ████ and uploaded to the common drive by Ms. ████ and decided against allowing me to compete for the promotion position.

The agency's actions undoubtedly had a chilling effect on the use of the EEO process at PEO STRI. Their actions also violated prohibitions against interfering with the EEO process. The agency's actions violated the rehabilitation act provisions regarding the confidentiality of employee medical history and/or records which includes medical information obtained from the employee.

I humbly request that the EEOC take appropriate actions that will ensure Ms. ███████ and Ms. ███ do not take similar detrimental actions in the future related to releasing employee personally identifiable and confidential information.

Respectfully submitted,

Appendix H

LETTER TO PRESIDENT OBAMA (2015)

December 7, 2015

President Barrack Obama The

White House

1600 Pennsylvania Avenue

Washington, D.C. 20500

Dear Mr. President,

My name is Harroll Ingram and I have served the taxpayers and service members for over 25 years as a DOD civilian (Engineer). I have had a great career in government, but hit some bumps once I joined the civilian Army workforce. I was forced to seek assistance from the MSPB and EEOC. The MSPB extended help in one case, but not in others. The EEOC left me feeling re- victimized. This letter is not really about me, but does concern the other people who serve as civilians and are forced to hope the MSPB and EEOC will aid them in correcting the wrongs of their leaders.

I was shocked to find that the MSPB has steadily held a less than 10% rate of assisting civilians who seek help concerning the improper actions of their agency leaders. For decades, by their own admission, the MSPB has ruled in favor of the agency more than 90% of the time concerning the thousands of cases they have adjudicated. That alone points to a likely issue relative to attaining the vision of the MSPB (Vision: A highly qualified, diverse Federal workforce that is fairly and effectively managed, providing excellent service to the American people).

The MSPB employees seem to be crying out for help in meeting their mission. According to a recent MSPB report dated May 29, 2015, "the proportion of MSPB employees who agreed they had the resources they needed rose. However, it was still lower than the proportion who disagreed that they had the resources they needed." Based on my experiences with the MSPB and EEOC, I am not certain additional employees and other resources will correct the problematic more than 90% rulings in favor of the agencies.

Mr. President; I humbly ask that you please consider assigning a team to review past MSPB decisions to include actually viewing the submitted evidence. The rulings the MSPB sent to me did not correctly identify and/or address the evidence submitted. One decision (Docket # AT- 1221-14-0725-W-1) grossly misspelled names in an effort to protect leaders and ignored submitted evidence. Knowing we have a group reviewing the decisions of the MSPB and EEOC will likely alleviate feelings of re-victimization held by the federal employees at the lower levels. It will also assist in eradicating any corruption that exists in the organizations relative to protecting certain leaders.

Most respectfully,

The Child Within the Scholar Recognizes Hypocrisy

www.ingramcontent.com/pod-product-compliance
Lightning Source LLC
Chambersburg PA
CBHW080622030426
42336CB00018B/3048